Palgrave Studies in Nineteenth-Century Writing and Culture

General Editor: **Joseph Bristow**, Professor of English, UCLA

Palgrave Studies in Nineteenth-Century Writing and Culture is a new monograph series that aims to represent the most innovative research on literary works that were produced in the English-speaking world from the time of the Napoleonic Wars to the *fin de siecle*. Attentive to the historical continuities between 'Romantic' and 'Victorian,' the series will feature studies that help scholarship to reassess the meaning of these terms during a century marked by diverse cultural, literary, and political movements. The main aim of the series is to look at the increasing influence of types of historicism on our understanding of literary forms and genres. It reflects the shift from critical theory to cultural history that has affected not only the period 1800–1900 but also every field within the discipline of English literature. All titles in the series seek to offer fresh critical perspectives and challenging readings of both canonical and non-canonical writings of this era.

Titles include:

Laurel Brake and Julie F. Codell (*editors*)
ENCOUNTERS IN THE VICTORIAN PRESS
Editors, Authors, Readers

Colette Colligan
THE TRAFFIC IN OBSCENITY FROM BYRON TO BEARDSLEY
Sexuality and Exoticism in Nineteenth-Century Print Culture

Dennis Denisoff
SEXUAL VISUALITY FROM LITERATURE TO FILM, 1850–1950

Laura E. Franey
VICTORIAN TRAVEL WRITING AND IMPERIAL VIOLENCE

Lawrence Frank
VICTORIAN DETECTIVE FICTION AND THE NATURE OF EVIDENCE
The Scientific Investigations of Poe, Dickens and Doyle

Jarlath Killeen
THE FAITHS OF OSCAR WILDE
Catholicism, Folklore and Ireland

Stephanie Kuduk Weiner
REPUBLICAN POLITICS AND ENGLISH POETRY, 1789–1874

Kirsten MacLeod
FICTIONS OF BRITISH DECADENCE
High Art, Popular Writing and the *Fin de Siècle*

Diana Maltz
BRITISH AESTHETICISM AND THE URBAN WORKING CLASSES, 1870–1900

Catherine Maxwell and Patricia Pulham (*editors*)
VERNON LEE
Decadence, Ethics, Aesthetics

David Payne
THE REENCHANTMENT OF NINETEENTH-CENTURY FICTION
Dickens, Thackeray, George Eliot and Serialization

Julia Reid
ROBERT LOUIS STEVENSON, SCIENCE, AND THE *FIN DE SIÈCLE*

Anne Stiles (*editor*)
NEUROLOGY AND LITERATURE, 1860–1920

Ana Parejo Vadillo
WOMEN POETS AND URBAN AESTHETICISM
Passengers of Modernity

Phyllis Weliver
THE MUSICAL CROWD IN ENGLISH FICTION, 1840–1910
Class, Culture and Nation

Palgrave Studies in Nineteenth-Century Writing and Culture
Series Standing Order ISBN 0-333-97700-9 (hardback)
(*outside North America only*)

You can receive future titles in this series as they are published by placing a standing order. Please contact your bookseller or, in case of difficulty, write to us at the address below with your name and address, the title of the series and the ISBN quoted above.

Customer Services Department, Macmillan Distribution Ltd, Houndmills, Basingstoke, Hampshire RG21 6XS, England

Neurology and Literature, 1860–1920

Edited by Anne Stiles

palgrave
macmillan

First published 2007 by
PALGRAVE MACMILLAN
Houndmills, Basingstoke, Hampshire RG21 6XS and
175 Fifth Avenue, New York, N.Y. 10010
Companies and representatives throughout the world

PALGRAVE MACMILLAN is the global academic imprint of the Palgrave Macmillan division of St. Martin's Press, LLC and of Palgrave Macmillan Ltd. Macmillan® is a registered trademark in the United States, United Kingdom and other countries. Palgrave is a registered trademark in the European Union and other countries.

ISBN-13: 978–0–230–52094–3 hardback
ISBN-10: 0–230–52094–4 hardback

This book is printed on paper suitable for recycling and made from fully managed and sustained forest sources. Logging, pulping and manufacturing processes are expected to conform to the environmental regulations of the country of origin.

A catalogue record for this book is available from the British Library.

Library of Congress Cataloging-in-Publication Data

Neurology and literature, 1860–1920/edited by Anne Stiles.
p. cm.
Includes bibliographical references (p.) and index.
ISBN 0–230–52094–4 (alk. paper)
1. English literature—19th century—History and criticism. 2. English literature—20th century—History and criticism. 3. Literature and science—Great Britain—History—19th century. 4. Literature and science—Great Britain—History—20th century. 5. Neurosciences—Great Britain—History—19th century. 6. Neurosciences—Great Britain—History—20th century. 7. Mind and body in literature. I. Stiles, Anne 1975–

PR468.S34N48 2007
820.9'356—dc22

2007016450

10 9 8 7 6 5 4 3 2 1
16 15 14 13 12 11 10 09 08 07

Printed and bound in Great Britain by
Antony Rowe Ltd, Chippenham and Eastbourne

Contents

IV. The Traumatized Brain

List of Illustrations

Acknowledgments

This volume was greatly enriched by the insights of participants in two conference panels on neurology and literature, the first at the Northeast Modern Language Association Conference in Philadelphia and the second at the American Comparative Literature Association Conference in Princeton, New Jersey, both of which took place in spring, 2006. I would like to express my gratitude to all of the panelists and enthusiastic audience members who made these sessions so rewarding.

Several institutions generously provided financial support during the compilation and editing of this volume. The English Department at the University of California, Los Angeles and the University of California Humanities Research Institute funded the early stages, while the American Academy of Arts and Sciences supplied me with office space and access to Harvard University's libraries during the final phase of revising and proofreading.

The cover artwork is included by kind permission of "Herederos de D. Santiago Ramón y Cajal" (Heirs of D. Santiago Ramón y Cajal). I am particularly grateful to Jorge Ramón y Cajal Asensio; Maria Angeles Ramón y Cajal; Russell Johnson at the Louise M. Darling Biomedical Library at the University of California, Los Angeles; and Javier DeFelipe at the Cajal Institute in Madrid for their assistance in this matter.

Thanks are also due to Paula Kennedy and the staff at Palgrave Macmillan, and to the anonymous reader who provided insightful feedback on each chapter. Finally, I am most grateful to series editor Joseph Bristow for his painstaking attention to this manuscript at every stage of its development, not to mention his enthusiastic encouragement along the way.

Anne Stiles
Washington State University

Notes on Contributors

James Kennaway is currently a postdoctoral research fellow in the music department at Stanford University. After studying at the London School of Economics, the Wellcome Institute for the History of Medicine, and King's College London, he completed a Ph.D. in Musicology on the subject of Wagner and degeneration at UCLA in 2004. In 2005–6 he taught at the University of Frankfurt-an-der-Oder in Germany.

Randall Knoper teaches English and American Studies at the University of Massachusetts Amherst. He is the author of *Acting Naturally: Mark Twain in the Culture of Performance* (University of California Press, 1995) and of various articles on U.S. literature and American Studies. He is currently writing a book on American literature and neuroscience with a focus on the late-nineteenth and early-twentieth centuries.

Don LaCoss teaches European and world history at the University of Wisconsin, LaCrosse. He has written a number of essays on the surrealist movement and its genealogies, and co-edited an anthology entitled *Surrealism, Politics and Culture* (Ashgate, 2003). He is currently researching the influence of German Romanticism, Symbolism, and Decadence on Arab surrealism in Cairo and Alexandria during the British military occupation of World War Two.

Andrew Mangham is Lecturer in English literature at the University of Reading. He is the author of *Violent Women and Sensation Fiction: Crime, Medicine, and Victorian Popular Culture* (2007) and editor of *Wilkie Collins: Interdisciplinary Essays* (2007).

Jill Matus is Professor of English and Vice-Principal of University College at the University of Toronto. She is the author of *Unstable Bodies: Victorian Representations of Sexuality and Maternity* (1995), *Toni Morrison* (1998), and the editor of *The Cambridge Companion to Elizabeth Gaskell* (2006). She has published many essays on the relations of Victorian literature to medical and psychological discourse of the period. Her current project, of which the chapter in this volume is a part, is a book-length study of psychic shock, which historicizes the emergence of trauma as a concept through a focus on Victorian theories of mind, consciousness, memory, and emotion in literary and psychological texts.

Mark S. Micale (Ph.D., Yale University, 1987) is Associate Professor of History at the University of Illinois in Champaign-Urbana where he teaches courses in the history of science and medicine, modern European culture and thought, and the history of France. He is the author or editor of several books, including *Beyond the Unconscious* (1993), *Discovering the History of Psychiatry* (1994), *Approaching Hysteria* (1995), *Enlightenment, Passion, Modernity* (2000), *Traumatic Pasts* (2001), and *The Mind of Modernism: Medicine, Psychology, and the Cultural Arts in Europe and America, 1870–1940* (2004). He is currently completing a study of medicine and masculinity from the Renaissance to Freud.

Laura Otis, who trained in Biochemistry and Neuroscience, is now Professor of English at Emory University, where she teaches courses on the relationship between literature and science. In 2000, she received a MacArthur Foundation grant to pursue interdisciplinary research. Otis is the author of *Organic Memory* (1994), *Membranes* (1999), and *Networking* (2001) and the translator of Santiago Ramón y Cajal's *Vacation Stories* (2001). Her latest book, *Müller's Lab* (2007) explores the history of science from multiple perspectives.

Anne Stiles is Assistant Professor of English at Washington State University in Pullman, where she teaches Victorian literature. Her monograph-in-progress, *Reading the Neurological Romance: Popular Fiction and Brain Science, 1865–1905,* explores rhetorical and philosophical connections between late-nineteenth-century neurology and works by Robert Louis Stevenson, Bram Stoker, Marie Corelli, and H.G. Wells. Drafts of two chapters have appeared in *Studies in English Literature, 1500–1900* and *Journal of the History of the Neurosciences.*

Kristine Swenson is Associate Professor of English at the University of Missouri-Rolla, where she teaches nineteenth- and twentieth-century literature. She recently published *Medical Women and Victorian Fiction* (University of Missouri Press, 2005), a monograph on representations of nurses and women doctors in the second half of the nineteenth century in Britain and the U.S. Her current book project is on the turn-of-the-century writers and sisters, Arabella and Annesley Kenealy.

Introduction

Anne Stiles

Neurology and literature are disciplines that initially appear to have little, if anything, to do with one another. The first is a so-called hard science practiced by a select coterie of medical doctors and researchers, while the second is a pleasurable artistic pursuit, theoretically open to all literate individuals. But first impressions can be deceptive. The present collection of essays aims to demonstrate that, in the late-nineteenth and early-twentieth centuries at least, brain science and imaginative fiction shared common philosophical concerns and rhetorical strategies.

The time period covered in this study has been delimited primarily by developments in neurological, rather than literary history – though sometimes these two fields serendipitously overlapped. Beginning in the early 1860s, neurology and the study of language collided dramatically when French neurologist Paul Broca (1824–1880) linked the third frontal convolution of the left brain hemisphere to linguistic ability.[1] Broca's findings immediately inspired his scientific peers to trace other mental faculties back to discrete cerebral locations, ushering in a period of biological determinism and physiological reductionism that reigned until shortly after the First World War, when Sigmund Freud's psychoanalytic approach gained broader currency throughout Europe and America. Freud began developing psychoanalysis in the 1890s when, despite his early training as a neurologist, he "gradually came round to the view . . . that psychical processes can only be dealt with in the language of psychology."[2] As Elaine Showalter and others have argued, Freud's psychoanalytic methods gained widespread popularity outside German-speaking nations only after the return of shell-shocked battle veterans suggested the need for new treatments.[3] During the six previous decades, therefore, biological explanations of psychological states held sway.

This era of biological reductionism coincided with a period of immense literary output, not to mention fruitful interdisciplinary collaboration between authors and neurologists. During this period, neurologists like Silas Weir Mitchell and Santiago Ramón y Cajal wrote moving fiction, while novelists like H.G. Wells, Robert Louis Stevenson, and Wilkie Collins penned literary works dramatizing neurological hypotheses and probing the philosophical ramifications of scientific discoveries. Clearly, scientists and artists of the 1860 to 1920 period were paying very close attention to one another. Indeed, the essays in the present volume emphasize how exchanges between literary and scientific writers during these six decades were not simply reflective – science influencing literature or vice versa – but rather dialogic or circular, a conversation where literary and scientific authors were mutually responsive to one another.

What we can learn from the interdisciplinary conversations taking place at this time is that the so-called two cultures problem described by C.P. Snow and F.R. Leavis in the mid-twentieth century was only beginning to take shape during the late-Victorian and early-Edwardian periods.[4] During this time, artists and scientists interacted on many levels, collaborating, quarreling, and generally struggling to find ideological common ground. Although these disparate groups of intellectuals often disagreed about the methods and ethical consequences of scientific inquiry, they shared a common bond arising from their ambivalence about the philosophical ramifications of scientific materialism and physiological reductionism.

Neurology, psychology, and the mind–body problem

Describing the state of psychology in the early 1890s, William James remarked, "Mental facts cannot be properly studied apart from [their] physical environment . . . the great fault of the older rational psychology was to set up the soul as an absolute spiritual being."[5] As James acknowledged, the men responsible for this materialist trend in late-Victorian mental science included neurologists like David Ferrier and John Hughlings Jackson in Britain, Paul Broca in France, and Gustav Fritsch and Eduard Hitzig in Germany. In scientific circles, these men were widely respected for their successful attempts to correlate specific mental and physical behaviors with discrete brain regions. Their experiments, especially Ferrier's, ushered in the modern era of neurosurgery, in which neurosurgeons could save lives by using functional maps of the brain to locate tumors, infections, and skull fractures.[6] To the general public and even conservative scientists, however, men such as Ferrier and Jackson often figured as villains, due to their controversial research methods (especially vivisection) and the obvious ways in which their research undermined the widespread lay perception of the "soul" or the "will" as the governing force behind human action.[7] Unlike René Descartes, whose *Traité de l'homme* (1664) confidently posited the pineal gland as the corporeal location of the soul, these late-Victorian researchers could conceive of no physical locus for spirituality in the human brain.

Ironically, Descartes himself was the source of this quandary. Since its inception in the second half of the seventeenth century, modern neurology has failed to come to terms with Cartesian mind–body dualism: that is, the idea of reality as a dichotomy between matter and spirit.[8] Though Descartes solved the problem to his own satisfaction with his theory about the pineal gland, his most enduring legacy, as psychologist Stanley Finger explains, was his "ideas about the brain as a reflexive machine."[9] As historian Robert Young argues, the mind–body problem "finds its most precise scientific expression in the related problems of classifying and localizing the functions of the brain."[10] In the minds of many laymen and even some scientists, including French physiologist Jean Pierre Marie Flourens, the question of pinpointing the cerebral origin of movements and thoughts apparently "undermin[ed] the unity of the soul, human immortality, free will, and the very existence of God."[11]

Because cerebral localization theories raised such disturbing doubts, much nineteenth-century neurological research initially met with ambivalent or openly hostile public reception. Phrenologist Franz Joseph Gall (1758–1828), one of the first scientists to argue that the brain was the seat of the emotions as well as the intellect, as well as (arguably) the first proponent of cerebral localization, found himself subjected to constant charges of materialism and fatalism, and was even refused a Christian burial by the Catholic Church as a result of his controversial researches.[12] David Ferrier (1843–1928), meanwhile, found not only his experimental methods but also his theoretical conclusions (and their philosophical ramifications) on trial when he was accused of violating the Anti-Vivisection Act in 1881. While Ferrier was eventually acquitted, a great many Victorians (including some fellow scientists) felt threatened by the climate of "extreme physiological reductionism" that his research had helped to usher in.[13] For instance, prominent physiologist William Benjamin Carpenter lamented that the neurological perspective of the brain as a soulless machine not only offended revered religious ideals, but also contradicted "the universal testimony of experience" which would tend to suggest "the conception of an Ego as something unconditioned by material states and physical forces."[14]

While there were definite philosophical continuities between Gall's research and Ferrier's, the two men worked with markedly different scientific methods and concepts. The half-century between Gall's landmark work, *Sur les fonctions du cerveau* (1822–5), and Ferrier's similarly titled classic, *The Functions of the Brain* (1876), witnessed the emergence of neurology as a discipline, related to but distinct from developing sciences like psychology, psychiatry, and the study of mental illness.[15] Though there was much overlap between these disciplines during the Victorian era, nineteenth-century scientists themselves recognized the ways in which these fields increasingly diverged. As German physiologist Ewald Hering wrote in 1887:

> The neurologist is thus placed between the physicist and the psychologist. The physicist considers the causal continuity of all material processes as the basis of his inquiry; the thoughtful psychologist looks for the laws of conscious life according to the rules of an inductive method and assumes the validity of an unalterable order.[16]

Hering's remarks suggest that nineteenth-century neurologists occupied a liminal position, confining themselves neither to the study of the tangible apparatus of the brain and nerves nor to intangible matters of the psyche. This generalization held true particularly for experimental neurologists and physiologists, many of whom (like Carpenter) ruminated about the philosophical implications of their discoveries.

In clinical practice, however, Victorian and modern neurologists have tended to occupy themselves primarily with "material processes," including organic conditions of the brain and nervous system (especially brain lesions and tumors). By contrast, psychiatrists and psychologists then and now treat aberrant mental states whose organic causes are unknown or uncertain. The caveat here is that an increasing number of diseases now fall under the umbrella of neurology, since neurologists today know more about organic causes of psychological disorders than their nineteenth-century predecessors. Schizophrenia and epilepsy, for example, were once classified as psychological ailments, but increasingly fell under purview of neurologists as their organic causes were uncovered.[17]

Neurology also differs from psychology in that the results of neurological experiments are generally quantifiable to a greater extent than the findings of psychologists. Young argues, for example, that

> [t]he most fundamental and perplexing problem in psychology has been, and remains, the lack of an agreed set of units for analysis

comparable to the elementary particles in physics and the periodic table of elements in chemistry.[18]

By contrast, neurologists have been able to measure electrical impulses sent to and emitted by various parts of the brain, a process greatly assisted by the invention of electroencephalography (EEG), or brain wave recording, in 1929 by German neurologist Hans Berger.[19] Neurology also boasts an elementary particle of sorts, the neuron: another discovery of the late-nineteenth century. Spanish neurologist Santiago Ramón y Cajal first argued that nerve cells were independent elements in 1889; his findings were affirmed and popularized by Wilhelm von Waldeyer, who coined the term "neuron" in 1891.[20]

The quantitative aspects of neurology, along with its highly specialized scientific jargon, have tended to make neurological writings less accessible to lay readers than the research of neighbor disciplines such as psychology and psychiatry. While it is true that scientific terminology and quantitative methods have infiltrated some types of psychological research, much psychological writing remains highly readable to non-experts. One might even argue with Young that "[m]ore than any other science, psychology is obliged to make sense to the layman, for its explanatory task is to make sense of the behavior of the layman."[21] Evidence for this claim exists in the proliferation of popular psychology books in the late-twentieth and early-twenty-first centuries. By contrast, few authors besides Oliver Sacks and Joseph LeDoux have written books about neurology intended primarily for popular audiences.[22]

By virtue of its relative inaccessibility, neurology has generally been perceived by the lay public as more intellectually threatening than its neighbor disciplines. This perception has been reinforced by the materialist implications of much neurological research. Given their exclusive focus on the brain, neurologists have tended to implicitly exclude the concept of the soul from their research. By contrast, psychology does not necessarily exclude this concept, and is etymologically connected to it ("psyche" is Greek for soul).

Despite its relative inaccessibility, neurology was arguably the most prestigious of the mental sciences during the period under discussion. Accordingly, neurologists enjoyed far greater medical and social status than any other class of mental health professionals.[23] The reasons for neurologists' relative prominence during this period were manifold. Most important, the prestige of nineteenth-century scientific advances in the fields of thermodynamics, astronomy, geology, and biology encouraged medical practitioners to seek physical explanations for psychological phenomena.[24] Though the study of mental illness was far from an exact science at this time or at any later period, mental health practitioners were encouraged by cerebral localization experiments that seemed to hold out the hope that all brain injuries and diseases might eventually be treatable once their organic causes were known. In general, then, neurologists blazed trails which psychiatrists and psychologists avidly followed, so that nineteenth- and early-twentieth century mental health practitioners generally favored somatic interpretations of mental phenomena.

Literary responses to neurological discoveries

Given the frequency with which neurological discoveries and methods appeared in the popular press, not to mention the social and intellectual prestige of neurology as a discipline, the Victorian public could hardly remain ignorant of the perceived threat imposed by a highly influential, increasingly professionalized science that challenged their most fundamental beliefs about their place in the universe. Authors of fiction, particularly those well trained in the sciences, were among the most articulate public figures to voice their concerns about new neurological developments. Many of these literary authors responded with reactionary fervor. Marie Corelli, by far the most commercially successful late-Victorian novelist (although not the best known to posterity), feared that "were science to unveil her marvels too openly to semi-educated and vulgarly constituted minds, the result would be, first Atheism, next Republicanism, and finally Anarchy and Ruin."[25] Like many late-Victorian authors, she worried that neurologists who cited biological origins for insane and criminal behaviors discouraged people from listening to the promptings of conscience. In her novel *Wormwood* (1890), the absinthe-addicted protagonist excuses his criminal actions on physiological grounds: "Plenty of scientists and physiologists could be found to prove that my faults are those of temperament and brain-construction, and that I cannot help them if I would."[26] One might also categorize Bram Stoker, the author of *Dracula* (1897), as a Luddite of sorts.[27] In his first novel, *The Snake's Pass* (1890), one character quips, "For real cold-blooded horror, commend me to your men of science." [28]

However, not all authors of fiction during this period reacted against recent neurological developments. Some novelists implicitly recognized (and readily capitalized upon) the similarities between, on the one hand, fictional works, and, on the other hand, scientific case studies as well as experimental narratives. Many of these scientific writings, in Stanley Finger's words, "could easily have come from an imaginative novelist."[29]

To name only one example, Robert Louis Stevenson readily adapts the genre of the case study for fictional purposes in his aptly titled novella, *Strange Case of Dr. Jekyll and Mr. Hyde* (1886).[30]

The great diversity of neurologically inflected British fiction during the 1860 to 1920 time period underscores that not all literary authors disapproved of recent neurological discoveries, despite their potentially disturbing philosophical implications. For instance, detective fiction by Wilkie Collins, Charles Dickens, and physician-author Sir Arthur Conan Doyle features protagonists who employ the empirical methods and objective world view of the natural sciences (including neurology) to solve mysterious crimes.[31] On the other side of the Atlantic, meanwhile, physician-authors like Silas Weir Mitchell and Oliver Wendell Holmes intervened in neurological debates by exploring the fascinating yet disturbing psychological consequences of nervous injury and abnormality in fictions like "The Case of George Dedlow" (1866), *Constance Trescot* (1905), and *A Mortal Antipathy* (1885). Around the same time, Continental writers as diverse as Friedrich Nietzsche, Arthur Schnitzler, Thomas Mann, Guy de Maupassant, and Emile Zola pondered the resonances of late-Victorian degeneration theories and the advent of Freudian psychoanalysis. Debates about neurology and its philosophical ramifications therefore likely reached almost every literate person in Europe and America. To some extent, fiction writers' responses to new discoveries can serve as an index of public reaction to these findings.

The essays in this volume examine the philosophical and cultural debates spawned by late-Victorian and Edwardian neurology as filtered through the lens of contemporary fiction, particularly in Anglophone nations but also in Continental Europe. This multinational focus reflects the cosmopolitan, international nature of the *fin-de-siècle* scientific community, and the ways in which neurological debates so frequently transcended national boundaries.[32] In a remarkably similar way, philosophical debates spawned by neurological discoveries transcended disciplinary and generic limitations, so that literary authors could contribute productively to conversations begun in scientific circles, and vice versa.

While this collection is hardly the first project to tackle the productive cultural interactions between literature and the mental sciences, it is, to my knowledge, the first book-length volume devoted specifically to the interactions between neurology and literature during the pivotal 1860 to 1920 period. Given the immense prestige of neurology and its practitioners, not to mention the disturbing philosophical implications of their research, it stands to reason that the relationship between neurology and literature during the 60 years under study would be more philosophically vexed and socially divisive than that between literature and psychology or psychiatry. This helps to explain why neurological controversies tended to surface most frequently in sensational or popular genres (the romance, the neo-Gothic novel, the detective novel, and the sensation novel) rather than high-cultural genres like realism, although there are certainly exceptions to this rule (the novels of George Eliot, for example).

The late-Victorian revival of the Gothic proved especially congenial to neurological debates, even though the worldview presented in Gothic fiction might initially seem at odds with late-nineteenth-century scientific rationalism. Kelly Hurley provocatively suggests that neo-Gothic fiction "should be seen as in opportunistic relation to the sciences," since the genre capitalizes on the "collective psychological demand" created by the traumatic impact of late-Victorian science on the lay public.[33] By contrast, Robert Mighall argues that "horror fiction has a generic obligation to evoke fear or suggest mystery," whereas "science . . . attempts to contain fear and offer a rational explanation for all phenomena."[34] It seems clear, then, that Victorian and Edwardian horror fiction utilized neurological facts and explanations both to provoke fear and to contain it. Accordingly, the essays in this volume address a wide range of literary authors and forms, but cluster particularly around a group of authors (including Wilkie Collins, H.G. Wells, and George Du Maurier) whose sensational works employed neurological concepts to play upon public fears of scientific materialism and physiological reductionism.

Overall, the critical approach applied in this volume can best be described as a synthesis of literary interpretation and the history of science. This collection builds upon interdisciplinary scholarship like Alan Richardson's seminal volume, *British Romanticism and the Science of the Mind* (2001) and George Rousseau's *Nervous Acts: Essays on Literature, Culture and Sensibility* (2004), both of which set important precedents by exploring similar concerns in the Romantic period and the Enlightenment, respectively. These two studies are unusual in that they explicitly address interactions between neurology and literature (as opposed to psychology and literature, madness and literature, etc.) while generally avoiding the approach typified by cognitive literary theory, a school of thought which anachronistically imposes twenty-first-century neurological concepts upon fictions of earlier time periods.[35] Because the four decades (1790–1830) covered by Richardson's study are chronologically and ideologically closer to the 1860 to 1920 period, his volume is of particular relevance here. Richardson emphasizes how Romantic mental sciences, no less than Victorian neurology, often called into question "the existence of the soul, the necessity of God, and the integrity of the self."[36]

Neurology and Literature attempts to emulate key features of Richardson's and Rousseau's works, but also to focus attention on a later, crucial period in the history of the neurosciences that differs from the Romantic and Enlightenment eras in important ways. As Edwin Clarke and L.S. Jacyna argue, the late-eighteenth and early-nineteenth centuries were characterized by a holistic view of nature as well as "a trend to search for synthesis, unity, and general laws in the life sciences rather than to concentrate solely upon narrowly conceived empirical studies and the accumulation of data for its own sake."[37] By contrast, late-nineteenth-century neurological studies became increasingly empirical and corporeally focused, leading to charges of physiological reductionism from within and without the scientific community.[38]

A second crucial difference between these periods was that mid-nineteenth-century evolutionary theories replaced the static conceptions of man's place in the universe that predominated during the Romantic era. During the late-eighteenth and early-nineteenth centuries, neurologists thought organic life existed along a "great chain of being" in which

> animate nature advanc[ed] through progressively more elaborate and elevated stages towards a preconceived goal. This pinnacle of perfection was the human body, but there was also hierarchy within the human frame . . . the nervous system represented the apogee of organic evolution; the point to which nature was striving, and to which all other systems of the body were subsidiary and preparatory.[39]

This relatively naive view of man's place in nature received its death-blow in 1859 with the publication of Darwin's *Origin of Species*. Darwinian evolutionary theory clearly demonstrated biological relationships between man and other animals, thereby implicitly challenging man's position at the "pinnacle" of the natural world. These crucial scientific and cultural differences between the two periods suggest that one cannot easily transpose previous findings about the relationship between Romantic neurology and literature onto the six decades under discussion here. The interactions between neurology and literature from 1860 to 1920 prove different enough to merit a separate study, particularly given the great significance of the neurological discoveries made during this time.

Late-Victorian and Edwardian interdisciplinarity

This volume aims, in part, to make a case for the 1860 to 1920 period as especially well suited to the kind of interdisciplinary inquiry undertaken here. In so doing, I must acknowledge those authors (such as Laura Otis,

Mark S. Micale, Rick Rylance, and Lilian R. Furst, among others) who have laid the groundwork for this project by exploring the interactions between science, psychology, and literature around the turn of the century, along with editors whose anthologies of Victorian scientific writing have made my task infinitely easier by providing access to primary sources that are often difficult to find.[40] While these authors do not specifically focus on interactions between neurology and literature, their scholarship has helpfully underscored the need to move beyond Freud in our examination of psychological and literary culture during the period under discussion. The immense popularity of Freud's psychoanalytic techniques in the twentieth century has partially obscured the degree to which physiological explanations of human behavior predominated during the late-Victorian and Edwardian periods, contributing to a materialist, biological-determinist cultural and intellectual climate.

The aforementioned critics have also emphasized that the 1860 to 1920 period witnessed immensely fruitful interdisciplinary collaborations, enabling the exchange of ideas between fields as disparate as neurology and literature. Rick Rylance describes the "generalist nature of Victorian intellectual culture," in which accomplished writers of all backgrounds "shared concern in an unfolding public network of debate over psychological problems."[41] Laura Otis goes further, maintaining that "[t]he notion of a 'split' between literature and science, of a 'gap' to be 'bridged' between the two, was never a nineteenth-century phenomenon."[42] Moreover, she suggests, "[s]cience was not perceived as being written in a 'foreign language' – a common complaint of twenty-first century readers . . . science was in effect a variety of literature."[43]

By 1900, emergent disciplines like psychology, neurology, and psychiatry had yet to develop highly specialized professional jargons, so that scientific articles in these fields were accessible to a general readership. This was, after all, a period during which Britain's leading philosophical journal, *Mind* (1876–present), frequently provided a venue for introducing the latest work in experimental psychology and neurology. Likewise, lay journals like *Fortnightly Review*, *The Nineteenth Century*, and *The Cornhill Magazine* in Britain and *The Century* and *Harper's New Monthly Magazine* in the United States regularly contained discussions of the latest scientific research penned by famous scientists. To cite only one example, Bordeaux physician Eugène Azam's famous case study of French multiple personality Félida X was first introduced to the British public in *Mind* in 1876, and featured shortly thereafter in *Cornhill*.[44]

Perhaps more tellingly, many late-Victorian and Edwardian intellectuals enjoyed careers spanning scientific and humanistic disciplines.

During this period, laypeople could and did intelligently intervene in scientific debates, commanding the attention and respect of scientific professionals. To name only two examples, one can cite the eclectic and extremely influential careers of Frances Power Cobbe and Frederic W.H. Myers. Like most Victorian women, Cobbe had little access to scientific education. Nonetheless, she wrote highly informed and frequently anthologized articles about dreams and unconscious cerebration, responding to theories of reflex action of the cerebrum advanced in the 1840s and 1850s by Thomas Laycock and William Benjamin Carpenter.[45] Cobbe was also an outspoken critic of scientific practices, particularly experiments involving vivisection. In 1876, Cobbe founded the highly influential Society for the Protection of Animals Liable to Vivisection (later known as the Victoria Street Society), the entity responsible for hauling neurologist David Ferrier into court on charges of animal cruelty.[46]

Myers, a classicist by training, devoted the latter part of his life to psychical research, co-founding the influential Society for Psychical Research in 1882. The Society sought to examine scientifically such *outré* phenomena as spirit rapping, haunted houses, death wraiths, and extrasensory perception. While such inquiries may seem highly unscientific to the twenty-first-century reader, Victorian examinations of occult and telepathic phenomena arguably fell within the limits of "legitimate" or mainstream science. At the height of its popularity, the Society's membership read like a *Who's Who* of *fin-de-siècle* psychology, boasting such distinguished members as Pierre Janet, Cesare Lombroso, G. Stanley Hall, and Sigmund Freud.[47] While laypeople could make such extensive contributions to scientific enterprises and debates, scientifically trained individuals could undertake artistic projects with seemingly little sense of incongruity. The enormously successful literary careers of practicing physicians like Arthur Conan Doyle, Oliver Wendell Holmes, Silas Weir Mitchell, and Arthur Schnitzler serve as cases in point.

The interdisciplinarity of the late-Victorian and Edwardian eras had much to do with educational and technological advances, as well as with the professional and social dimensions of science at this period. Mark S. Micale has emphasized that late-nineteenth-century standardized public education gave physicians and novelists common intellectual ground, while technological advancements increased the possibility for cross-disciplinary and cross-cultural communication and travel.[48] Moreover, at a time before institutional funding for scientific pursuits was widely available, scientists who were not independently wealthy had to be willing and able to support themselves through teaching, journalism, medical practice, writing, or public lecturing. Since the nineteenth-century scientist was expected to fit the mold of the gentleman scholar, a well-rounded, classical education, including a thorough knowledge of ancient and modern literature, was considered essential to his social and professional advancement. Nineteenth-century scientific articles themselves were often peppered with references to poetry and fiction, in order to establish a connection with a broadly educated readership. Simply put, "[f]or aspiring scientists, it paid to be well read."[49]

Given the open lines of communication between science and the arts at this time, it should come as no surprise that neurologists and literary authors shared rhetorical strategies in addition to common cultural reference points. While late-Victorian realist and naturalist novelists like Emile Zola and George Moore employed the scientific method in their minute observations of daily life, medical writing took a decidedly narrative turn with longer "novelistic" case studies, culminating in Freud's extensive explorations of his patients' subjective experience. In *Studies on Hysteria* (1895), Freud ruefully admitted that his narrative case studies lack "the serious stamp of science" because of their unintentional resemblance to fiction: "It still strikes me myself as strange that the case histories I write should read like short stories."[50]

At times, the resemblances between case studies and literary works were striking enough to obscure the line between fact and fiction. French psychologist Theodore Flournoy's best-selling case study of a patient with multiple personality disorder, *From India to the Planet Mars* (1900), was read both as a case study and as a novel when it first appeared. The work depicts the imaginary experiences of the gifted medium Hélène Smith, who "claimed under hypnosis to be the reincarnation of a Hindu princess . . . and of an inhabitant of the planet Mars, whose language she claimed to speak and whose landscapes she painted."[51] Perhaps more impressively, Silas Weir Mitchell's fictional account of an amputee suffering from phantom limb syndrome, presented in the short story "The Case of George Dedlow" (1866), was taken for reality by many who read the tale in *The Atlantic Monthly*. Mitchell recorded his surprise when "[the story] was at once accepted by many as the description of a real case. Money was collected in several places to assist the unfortunate man, and benevolent persons went to 'The Stump Hospital,' in Philadelphia, to see the sufferer and offer him aid."[52]

Just as fictional case studies were sometimes mistaken for reality, real case studies often profoundly influenced authors of fiction. For instance, Eugène Azam's aforementioned patient, Félida X, impressed scientific and lay readers with the possibility of numerous interior lives. Fittingly, the concept of multiple personalities was soon after fictionalized in Stevenson's *Strange Case of Dr. Jekyll and Mr. Hyde*, where the fractured protagonist ruminates, "Man will be ultimately known for a mere polity of multifarious, incongruous, and independent denizens."[53] Stevenson's fictional "case" in turn probably colored scientific work on multiple personality disorder written during the 1880s and 1890s by Frederic Myers and Scottish psychiatrist Lewis Bruce.[54] This example highlights why no simple reflective model of influence (science influencing literature or vice versa) will suffice for this volume; instead, one must speak of two-way conversations between disciplines or of broader cultural movements evolving out of several fields simultaneously.

Neurologists on mental illness

Among the most significant developments to evolve out of this interdisciplinary cultural milieu was a biologically based conception of mental illness. Scientists' attempts to localize functions in specific regions of the brain suggested the possibility that every mental illness might be traced back to a cerebral lesion. Even if doctors could not always locate the lesions allegedly responsible for their patients' madness, they expected that further research would enlighten them regarding the somatic origins of various psychological disorders.

In an effort to bring more psychological conditions under the umbrella of neurology, scientists sought new explanations for ancient diseases like epilepsy and hysteria. For instance, French neurologist Jean-Martin Charcot argued in the 1870s that hysteria had neurological, not gynecological, origins. This initially controversial suggestion was, nonetheless, widely accepted by the end of the nineteenth century, when the idea of the "wandering womb" as the origin of hysteria had largely been replaced by the idea of a dynamic or generalized "cerebral or spinal lesion" as the most likely cause of the disorder.[55] The newer theory had the advantage of being impossible to disprove experimentally.[56] Meanwhile epilepsy, a disease known since ancient times and formerly attributed to demon possession, was newly ascribed to neurological causes. British neurologist John Hughlings Jackson traced epileptic seizures back to a discharging lesion in the brain containing diseased cells that act madly during an epileptic attack. Jackson's groundbreaking findings were confirmed by twentieth-century EEG measurements, which suggested that the "the electricity in the brain of the epileptic discharges in precisely the way Jackson had suggested."[57]

During this period, scientists also explored the neurological origins of newer diseases like neurasthenia, obsessive-compulsive disorders, and railway spine. Neurasthenia or nervous exhaustion, a diagnosis that enjoyed immense popularity on both sides of the Atlantic until well into the twentieth century, was first described by American neurologists George Beard and Silas Weir Mitchell in the late 1860s and early 1870s.[58] In *American Nervousness: Its Causes and Consequences* (1881), Beard explains that the condition originates from "a narrow margin of nerve force," so that the sufferer's nervous system can be likened to "an electric battery that does not supply very much more electric force than is needful for [its] use."[59] The popularity of neurasthenia as a diagnosis stemmed not only from its relatively benign somatic origins, but also from the lack of social stigma attached to the condition and its relative usefulness as an umbrella diagnosis encompassing diverse symptoms like depression, anxiety, indecision, irritability, dyspepsia, insomnia, migraine, etc.[60]

The conditions now referred to as obsessive-compulsive disorder and post-traumatic stress disorder (PTSD) also trace their nosological roots back to the nineteenth century, although these illnesses were then known by different names. As Andrew Mangham argues in this volume, obsessive-compulsive disorder developed out of the nineteenth-century preoccupation with monomania. Monomania, too, could be traced back specifically to the brain, this time to a lesion of the will or intelligence.[61] What we now term PTSD also began as a preoccupation of nineteenth-century scientists like Herbert Page in England, Hermann Oppenheim in Germany, and Charcot in France, who debated whether or not the psychological distress of railway accident victims was caused by "traumatic neurosis" or physical injury to the brain or spinal column.[62] Well before World War One and the advent of "shell-shock," scientists treating the veterans of the Crimean, Franco-Prussian, and American Civil wars played an important role in the early history of trauma. While Jill Matus's essay in this volume examines discourses of mental shock arising from the Crimean and Franco-Prussian wars, Mark S. Micale's chapter explores physicians' reactions to post-traumatic stress in American Civil War veterans.[63] In general, the essays in this volume focus on *fin-de-siècle* neurological explanations for psychological ills rather than anachronistically applying modern diagnostic criteria, while acknowledging the historical continuities between Victorian and modern diagnostic categories.

Twenty-first-century perspectives

Since modern critics live in a post-Freudian era dominated by biological explanations for mental illness, we are uniquely qualified to comment upon the science of the 1860 to 1920 period, which featured similarly biological determinist views. The contributors to this volume come from varied backgrounds, including literary scholars, historians, and even musicologists, many of whom have extensive training in science and the history of science. All of us, however, share an interest in the emergent "interdiscipline" of neurology and literature, and explore ways in which these apparently disparate subjects intersected during this 60-year period.

The eight chapters that make up this volume are divided into four thematically interconnected sections. The essays in Part I, "Catalysts," discuss watershed events that drew public attention to neurological concepts. Laura Otis's essay, "Howled out of the Country: Wilkie Collins and H.G. Wells Retry David Ferrier," explores how authors of popular fiction represented the infamous Ferrier trial of 1881. This trial is significant not just for its notoriety, but also because it disseminated theories of cerebral localization to the general public, laying bare the tensions dividing late-Victorian scientific and popular culture. In response to the trial, novelists created unflattering literary caricatures of vivisectors, such as the eponymous villain of H.G. Wells's *Island of Dr. Moreau* (1896) and the nefarious neurologist Dr. Benjulia of Wilkie Collins's *Heart and Science* (1883).[64] Ultimately, Otis argues that these novels, like the trial itself, were fictional "events" that disseminated information about animal experimentation and neurology to the reading public.

In "Our Lady of Darkness: Decadent Arts and the Magnetic Sleep of Magdeleine G.," Don LaCoss describes a series of remarkable performances in 1904 by hypnotized dancer Magdeleine Guipet, whose seven-month tour of the opera houses and theater stages of Munich, Vienna, Stuttgart, Paris, and London was an immense critical and popular success. Before she became famous, Guipet had sought treatment for persistent headaches from Émile Magnin, professor of magnetopathy at the Paris École de Magnetisme. Magnin, who thought the young woman would benefit from biomagnetic trance therapy, became increasingly impressed by his patient's rhythmic movements under hypnosis when she heard music or commanding voices. He displayed his gifted patient, who claimed to have no formal dance training, before distinguished medical colleagues, inciting debates about animal magnetism, automatic behaviors, the origins of creativity, and unconscious cerebral functions. These medical debates resonated with Decadent artists and literary authors who had been exploring their own unconscious impulses through their visual and narrative creations.

Part II, "Diagnostic Categories," explores the shifts in diagnosis and treatment of mental illness that followed in the wake of late-nineteenth-century neurological discoveries. Andrew Mangham's essay, "How do

I Look? Dysmorphophobia and Obsession at the *Fin de Siècle,"* probes the history of obsessive-compulsive disorders, particularly body dysmorphic disorder. These conditions trace their nosological roots to the nineteenth century. In the 1830s, Jean Esquirol, Pinel's successor at the Salpêtrière, defined "monomania" as a form of mental alienation in which the sufferer is obsessed with a single object. In 1891, Italian psychologist Enrico Morselli coined the term "dysmorphophobia" to denote unhealthy fixation on perceived physical flaws. The concept of body obsession gradually migrated into late-Victorian fictions like Fyodor Dostoyevsky's *Notes from the Underground* (1864), Wilkie Collins's *The Law and the Lady* (1875), and Oscar Wilde's *Picture of Dorian Gray* (1890, revised 1891). Mangham relates how dysmorphophobic characters in these novels ultimately endanger themselves and others in an effort to become physically inconspicuous, prefiguring the twenty-first-century practice of plastic surgery.

The new diagnostic categories of neurasthenia and male hysteria feature prominently in Kristine Swenson's essay, "Doctor Zay and Dr. Mitchell: Elizabeth Stuart Phelps's Feminist Response to Mainstream Neurology." Phelps, a best-selling American novelist during the post–Civil War era, corresponded with Silas Weir Mitchell about her own neurasthenic symptoms, but found his theories and treatments inadequate. She wrote *Doctor Zay* in 1882 to demonstrate the virtues of female physicians and homeopathic medicine, both of which Mitchell deplored. The novel's protagonist, homeopath and New Woman Dr. Atalanta Zay, cures and later marries the male neurasthenic Waldo Yorke, who is a typical example of the "over-civilized," nervous, urban male described by Mitchell and George Beard in their writings on neurasthenia. Although the romance between the New Man and the New Woman was a common *fin-de-siècle* plot element, Phelps's novel was unique because of its medical context, and because of the female protagonist's active role in shaping her future husband. By creating a successful female physician protagonist with an unorthodox cure for male nervousness, Phelps defied the masculinist neurological tradition represented by doctors like Mitchell.

Part III, "Sex and the Brain," explores how literary works mediate late-nineteenth- and early-twentieth-century neurological theories about gender differences and sexual practices.

In "Trauma and Sexual Inversion, circa 1885: Oliver Wendell Holmes's *A Mortal Antipathy* and Maladies of Representation," Randall Knoper argues that novelist and Harvard physiology professor Oliver Wendell Holmes made a connection between trauma and sexual inversion at least a decade before Sigmund Freud and Josef Breuer's *Studies in*

Hysteria (1895) famously suggested a causal relationship between childhood sexual trauma and neurosis. While still an infant, the protagonist of *A Mortal Antipathy* (1885), Maurice Kirkwood, is dropped off a balcony into a thornbush by his beautiful and nubile cousin. As a result, he experiences "inverted" nervous energies that threaten to stop his heart whenever he sees a sexually attractive young woman. Though Kirkwood does not experience same-sex desire, heterosexual arousal is potentially fatal for him. Kirkwood's unusual disorder reflects contemporary discourses surrounding nerve force that described "antipathetic instinct" as the opposite of magnetic attraction. Holmes's depiction of Kirkwood's ailment also invokes nineteenth-century debates about male hysteria and railway spine, along with contemporary sexological theories that included homosexuality in their catalog of sexual dysfunctions.

In "Singing the Body Electric: Nervous Music and Sexuality in *Fin-de-Siècle* Literature," James Kennaway explores why late-nineteenth-century scientists viewed certain kinds of music as pathogenic, especially for inherently "nervous" populations like women and homosexuals. For instance, George Beard's descriptions of neurasthenia emphasized the potentially overstimulating effects of music on those with "weak nerves," while Canadian scientist and novelist Grant Allen warned that auditory nerves could be "wearied by over-use or jarred by discordant sounds"[65] Music by composers of the so-called *neudeutsche Schule*, including Richard Wagner, Franz Liszt, and Hector Berlioz, was considered particularly harmful to the nervous system because the lush instrumentation and daring harmonies allegedly appealed to the senses rather than to the rational mind, provoking erotic and neurotic responses. Literary authors depicted the dangers of nervous music in works as diverse as Thomas Mann's *Wälsungenblut* (1905) and *Buddenbrooks* (1901), Joris-Karl Huysmans's *À rebours* (1884), and George Du Maurier's *Trilby* (1894). In general, novelists either viewed appreciation of nervous music as a sign of aristocratic sensibility, or denounced such music for inciting degeneracy and sexual deviance.

The fourth and final segment of this volume, "The Traumatized Brain," focuses on late-Victorian theories of traumatic neurosis that paved the way for the modern diagnosis of PTSD. In "Emergent Theories of Victorian Mind Shock: From War and Railway Accident to Nerves, Electricity and Emotion," Jill Matus reminds us that trauma is a historically produced concept rather than a timeless and transcendent one, and that the genesis of modern trauma theory pre-dates Freudian psychoanalysis. Drawing on examples from George Eliot's *Middlemarch* (1871–2), medical accounts of war and railway disasters in the 1860s and 1870s, and the physiological psychology of Herbert Spencer, Alexander Bain, and George Henry Lewes, Matus explores the contexts in which a discourse of psychic shock emerged in the mid- to late-nineteenth century. These Victorian discourses of psychic shock, particularly the effects of emotion on consciousness, provide a genealogy that may help to broaden the rather narrow focus on memory in twenty-first-century trauma theory.

Mark S. Micale's essay, "Medical and Literary Discourses of Trauma in the Age of the American Civil War," explores the reactions of Civil War physicians to the post-traumatic stress exhibited by their patients. Focusing primarily on the medical writings and quasi-autobiographical fiction of Silas Weir Mitchell, Micale explains why Civil War physicians generally failed to understand the psychological complaints of the battle-fatigued veterans they encountered in their practices. Civil War physicians did not yet have the social or conceptual framework necessary to formulate a theory of battle-induced psychological trauma. Thus, they tended either to focus exclusively on physical symptoms of emotional shock, or to condemn battle-fatigued soldiers as deserters and malingerers. Only with the advent of Freudian psychoanalysis and its application during World War One did wartime trauma, first labeled "shell-shock" and later "Post Traumatic Stress Disorder," enter the existing repertoire of medical diagnoses.

None of these sections is exhaustive, nor is this volume meant to be the last word on the subject of neurology and literature. Instead, this collection is intended to raise provocative questions and promote future scholarship in this interdisciplinary subject area. Many of the questions raised by a study of neurology and literature during the 1860 to 1920 period shed light on debates that are very much alive in the early-twenty-first century, for instance: What is the role of the soul or will in physiological conceptions of the human organism? Can scientific materialism coexist with a belief in free will or faith in a higher power? Is mental disease biological in origin, or determined by environment? Are there innate differences between the brains of men and women, heterosexuals and homosexuals, and individuals of diverse racial origins, and if so, how should social and medical policy makers use this information? What animal and human sacrifices are allowable in the name of scientific progress? And finally, perhaps most importantly, is it possible to maintain a stable identity or "self" when contemporary scientific theory defines us as the sum of our neurochemical impulses?

The time is right for such a study, since the end of the nineteenth century and the close of the twentieth have proven remarkably similar in several respects. Both the late-nineteenth and the late-twentieth centuries privileged somatic approaches to treating mental illness, whereas the beginnings of the twentieth and twenty-first centuries provoked introspection about social and environmental causes of psychological dysfunction (Freud's theories of psychoanalysis serve as the most obvious case in point). In 1994, historians of science Roy Porter and Mark S. Micale suggested that "Contemporary Western psychiatry is experiencing a 'paradigm shift' as it progressively deprivileges psychodynamic psychologies and embraces new, more biologically oriented theories and therapies."[66] A little over ten years later, there are signs that the pendulum may swing again, as patients gradually turn from primarily somatic therapies like medication toward holistic therapies (including, but not limited to acupuncture, biofeedback, meditation, self-help groups, and religious or spiritual philosophies). The increased popularity of mind– body approaches to healing among medical professionals presents another indication of this sea change.[67] Although it is too soon to say how extensively these trends will undermine the primarily somatic bias of contemporary medicine, these alterations in medical theory and practice resemble the shift toward psychoanalysis at the beginning of the twentieth century. Examining the interactions between neurology and literature at the turn of the twentieth century can offer much insight into the medical paradigm shift potentially looming on our horizon.

Notes

1. Anne Harrington, *Medicine, Mind, and the Double Brain: A Study in Nineteenth-Century Thought* (Princeton, NJ: Princeton University Press, 1987), 40–5.
2. James Strachey, "Editor's Introduction" to volume two of Sigmund Freud, *The Standard Edition of the Complete Psychological Works of Sigmund Freud*, ed. James Strachey and Anna Freud, 24 vols (1955; London: Hogarth Press, 1981), 2: xxiv. Strachey adds, "To the end of his life, however, Freud continued to adhere to the chemical aetiology of the 'actual' neuroses and to believe that a physical basis for all mental phenomena might ultimately be found" (2: xxiv). For a more detailed account of Freud's neurological training, particularly the influence of his mentor Theodor Meynert, see Peter Amacher, "Freud's Neurological Education and its Influence on Psychoanalytic Theory," *Psychological Issues* 4.4 (1965), 21–8.
3. Elaine Showalter, *The Female Malady: Women, Madness, and English Culture,* 1830–1980 (New York: Pantheon Books, 1985), 189–90.
4. See C.P. Snow, *The Two Cultures and a Second Look* (New York: Cambridge University Press, 1979); and F.R. Leavis, *Two Cultures? The Significance of C.P. Snow* (London: Chatto and Windus, 1962).
5. William James, *Psychology* (New York: Henry Holt, 1908), 3.
6. Stanley Finger, *Origins of Neuroscience: A History of Explorations into Brain Function* (New York: Oxford University Press, 1994), 41, 438–9.

7. On Ferrier's highly public trial for animal cruelty and the negative press coverage of his research, see Laura Otis in this volume, as well as Richard French, *Antivivisection and Medical Science in Victorian Society* (Princeton: Princeton University Press, 1975), 200–2 and 310–3. For a more general discussion of the ways in which neurologists and physiologists were vilified in *fin-de-siècle* fiction and popular press, see Coral Lansbury, *The Old Brown Dog: Women, Workers and Vivisection in Edwardian England* (Madison: University of Wisconsin Press, 1985).
8. The conceptual beginnings of what may be termed modern (as opposed to ancient or classical) neurology are open to debate. Etymologically, however, one can determine that Thomas Willis "coined the term *neurologie* in the 1660s; during the subsequent century, it assumed its current form and by the 1830s had spawned the term *neurologist* to signify someone who made nerves his special subject of inquiry." Janet Oppenheim, *"Shattered Nerves": Doctors, Patients, and Depression in Victorian England* (New York: Oxford University Press, 1991), 28.
9. Finger, *Origins of Neuroscience*, 26.
10. Robert Young, *Mind, Brain, and Adaptation in the Nineteenth Century: Cerebral Localization and its Biological Context from Gall to Ferrier* (New York: Oxford University Press, 1990), xxii.
11. Jean Pierre Marie Flourens, *Phrenology Examined* (1846), paraphrased in Harrington, *Medicine, Mind, and the Double Brain*, 9.
12. Young, *Mind, Brain, and Adaptation*, 16; Harrington, *Medicine, Mind, and the Double Brain*, 9n.
13. Young, *Mind, Brain, and Adaptation*, 26.
14. William Benjamin Carpenter, "On the Doctrine of Human Automatism: Part II," *Contemporary Review* 25 (1875), 941, 943. Carpenter famously invented the theory of "unconscious cerebration" that seemed to suggest that human beings were automata without free will. As one can see from this quotation, however, Carpenter was ambivalent about the materialist implications of his own discovery.
15. These dates are admittedly approximate. I base my claim here upon a related argument by Young: "In the sixty years between the publication of Gall's work on *The Functions of the Brain* and the appearance of Ferrier's volume of the same name . . . the study of the brain and its functions had become an experimental science based on the theory of evolution" (*Mind, Brain, and Adaptation*, 7).
16. Ewald Hering, "Memory as a General Function of Organized Matter," *The Open Court, a Quarterly Magazine* 1.6 (April 1887), IIB.
17. Thanks are due to neurologist Peter J. Koehler of Atrium Medical Centre in Heerlen, the Netherlands for many of the insights in this paragraph. Peter J. Koehler, letter to the author, 18 August 2006.
18. Young, *Mind, Brain, and Adaptation*, xx.
19. Finger, *Origins of Neuroscience*, 42, 251.
20. Finger, *Origins of Neuroscience*, 47.
21. Young, *Mind, Brain, and Adaptation*, xix.
22. See, for example, Sacks's eminently readable works such as *The Man Who Mistook His Wife for a Hat and Other Clinical Tales* (1970), or his autobiographical "neurological novel," *A Leg to Stand On* (1984). His works demonstrate an intriguing tendency to straddle the boundary between case study

and fiction. LeDoux's *The Emotional Brain: The Mysterious Underpinnings of Emotional Life* (1996) is a more recent popular and accessible work relating to neurology.

23. See chapters 1 and 2 of Oppenheim, *Shattered Nerves*, for a fuller elucidation of the social and professional hierarchies that privileged neurologists above other mental health practitioners.
24. Oppenheim, *Shattered Nerves*, 37.
25. Marie Corelli, *Ardath: The Story of a Dead Self* (London: Methuen, 1907), 141. One of Corelli's recent biographers, Brian Masters, describes the extent of her popularity: "While Queen Victoria was alive, Miss Corelli was the second most famous Englishwoman in the world . . . At least half of her thirty books were bestsellers. About 100,000 copies were sold every year. Her nearest rival was Hall Caine, who sold approximately 45,000 a year." *Now Barrabas was a Rotter: The Extraordinary Life of Marie Corelli* (London: Hamish Hamilton, 1978), 6.
26. Marie Corelli, *Wormwood: A Drama of Paris* (Chicago: M.A. Donohue, 1890), 14.
27. On Stoker's *Dracula* as a reaction against late-Victorian neuroscientific developments, see Anne Stiles, "Cerebral Automatism, the Brain, and the Soul in Bram Stoker's *Dracula*," *Journal of the History of the Neurosciences* 15.2 (June 2006), 131–52.
28. Bram Stoker, *The Snake's Pass* (Dingle, Ireland: Brandon 1990), 59.
29. Finger, *Origins of Neuroscience*, 434.
30. On the similarities between *Jekyll and Hyde* and late-Victorian case studies, see Anne Stiles, "Robert Louis Stevenson's *Jekyll and Hyde* and the Double Brain," *Studies in English Literature, 1500–1900* 46.4 (Autumn 2006), 879–900.
31. On the epistemological similarities between detective fiction and natural science, see Lawrence Frank, *Victorian Detective Fiction and the Nature of Evidence: The Scientific Investigations of Poe, Dickens, and Doyle* (New York: Palgrave Macmillan, 2003).
32. One might even say that collaborations as well as arguments between neurologists in different nations, who spoke different languages, were the rule rather than the exception. Examples include the collaboration between Spaniard Santiago Ramón y Cajal and German scientists Wilhelm von Waldeyer and Rudolph Albert von Kölliker, not to mention the public debate between Friedrich Goltz and David Ferrier at the 1881 International Medical Congress in London. Finger, *Origins of Neuroscience*, 44–7, 54.
33. Kelly Hurley, *The Gothic Body: Sexuality, Materialism, and Degeneration at the* fin de siècle (New York: Cambridge University Press, 1996), 5–6.
34. Robert Mighall, *A Geography of Victorian Gothic Fiction* (New York: Oxford University Press, 1999), xxiv.
35. For examples of recent work in cognitive literary theory, see Alan Richardson and Ellen Spolsky, eds, *The Work of Fiction: Cognition, Culture and Complexity* (Burlington, VT: Ashgate, 2004), especially the bibliography appended to Richardson's introduction (26–9). Richardson's work in this volume is quite different from his more historically oriented *British Romanticism and the Science of the Mind*.
36. Alan Richardson, *British Romanticism and the Science of the Mind* (New York: Cambridge University Press, 2001), 12.
37. Edwin Clarke and L.S. Jacyna, *Nineteenth-Century Origins of Neuroscientific Concepts* (Berkeley: University of California Press, 1987), 1–2.

38. Young argues, for example, that Victorians' "simple application of the ideas of corpuscular matter and motion to the study of mind, brain, and adaptation impoverished psychology" (*Mind, Brain, and Adaptation*, 8).
39. Clarke and Jacyna, *Nineteenth-Century Origins of Neuroscientific Concepts*, 5–6.
40. See particularly Mark S. Micale, ed., *The Mind of Modernism: Medicine, Psychology, and the Cultural Arts in Europe and America, 1880–1940* (Stanford: Stanford University Press, 2004); Laura Otis, *Membranes: Metaphors of Invasion in Nineteenth-Century Literature, Science and Politics* (Baltimore: Johns Hopkins University Press, 1999) and *Networking: Communicating with Bodies and Machines in the Nineteenth Century* (Ann Arbor: The University of Michigan Press, 2001); Rick Rylance, *Victorian Psychology and British Culture, 1850–1880* (New York: Oxford University Press, 2000); Lilian R. Furst, *Between Doctors and Patients: The Changing Balance of Power* (Charlottesville: University of Virginia Press, 1998); and Ian Hacking, *Rewriting the Soul: Multiple Personality and the Sciences of Memory* (Princeton, NJ: Princeton University Press, 1995). Recent anthologies of Victorian psychological writing include Laura Otis, ed., *Literature and Science in the Nineteenth Century* (New York: Oxford University Press, 2002); Lilian R. Furst, ed., *Medical Progress and Social Reality: A Reader in Nineteenth-Century Medicine and Literature* (Albany: State University of New York Press, 2000); and Jenny Bourne Taylor and Sally Shuttleworth, eds, *Embodied Selves: An Anthology of Psychological Texts, 1830–1890* (New York: Oxford University Press, 1998).
41. Rylance, *Victorian Psychology and British Culture*, 1, 3.
42. Otis, Introduction to *Literature and Science in the Nineteenth Century*, xvii.
43. Otis, Introduction to *Literature and Science in the Nineteenth Century*, xvii.
44. Unsigned review of Eugène Azam, "Double Consciousness with Periodic Loss of Memory (Amnesia)," *Mind* 1.3 (July 1876), 414–6; Richard Proctor, "Dual Consciousness," *The Cornhill Magazine* 35 (January 1877), 86–105.
45. Some of Cobbe's essays on unconscious cerebration and dreams, as well as her anti-vivisection articles, are reprinted in Taylor and Shuttleworth's *Embodied Selves* and Otis's *Literature and Science in the Nineteenth Century*. Many of Cobbe's essays first appeared in popular periodicals like *The Fortnightly Review, Macmillan's Magazine*, and *The Galaxy*, undoubtedly reaching a wide audience.
46. French, *Antivivisection and Medical Science in Victorian Society*, 112–9. See also Otis in this volume.
47. "Members and Associates: March, 1907," *Proceedings of the Society for Psychical Research* 19 (1907), 463.
48. Micale, Introduction to *The Mind of Modernism*, 4–7.
49. Otis, Introduction to *Literature and Science in the Nineteenth Century*, xx.
50. Sigmund Freud, "Fräulein Elisabeth von R." in Josef Breuer and Sigmund Freud, *Studies on Hysteria*, trans. and ed. James Strachey and Anna Freud (New York: Basic Books, 1957), 160.
51. Micale, Introduction to *The Mind of Modernism*, 13.
52. Silas Weir Mitchell, Introduction to *The Autobiography of a Quack* and *The Case of George Dedlow* (New York: The Century, 1900), ix–x.
53. Robert Louis Stevenson, *Strange Case of Dr. Jekyll and Mr. Hyde*, ed. Katherine Linehan (New York: Norton, 2003), 48.
54. Shortly after the publication of Jekyll and Hyde, Myers's article "Multiplex Personality" appeared in *The Nineteenth Century* 20 (November 1886), 648–66.

Regarding Stevenson's potential influence on Lewis Bruce, see Finger, *Origins of Neuroscience*, 396–7.

55. Mark S. Micale, *Approaching Hysteria: Disease and Its Interpretations* (Princeton: Princeton University Press, 1995), 111.
56. Charcot was not the first to propose moveable, nonfixed lesions as potential sources of mental disease. Jacques Moreau suggested this idea in his earlier work, *La psychologie morbide dans ses rapports avec la philosophie de l'histoire ou de l'influence des névropathies sur le dynamisme intellectuel* (Paris: Librairie Victor Masson, 1859), 38n.
57. George Frederick Drinka, *The Birth of Neurosis: Myth, Malady, and the Victorians* (New York: Simon and Schuster, 1984), 105.
58. On Beard's and Mitchell's roles in the "discovery" of neurasthenia, see Tom Lutz, "Varieties of Medical Experience: Doctors and Patients, Psyche and Soma in America," in *Cultures of Neurasthenia from Beard to the First World War*, ed. Marijke Gijswijt-Hofstra and Roy Porter (New York: Rodolpi, 2001), 51–76.
59. George Beard, *American Nervousness: Its Causes and Consequences* (1881; New York: Arno Press, 1972), 9, 10.
60. For an exhaustive list of neurasthenic symptoms, see Beard, *American Nervousness*, 7–8.
61. Jan Goldstein, *Console and Classify: The French Psychiatric Profession in the Nineteenth Century* (New York: Cambridge University Press, 1987), 171.
62. On "railway spine," see Randall Knoper and Jill Matus in this volume, along with Drinka, *The Birth of Neurosis*, 108–22.
63. For more information on pre-Freudian trauma theory, see also Ruth Leys, *Trauma, A Genealogy* (Chicago: University of Chicago Press, 2000); and Mark S. Micale and Paul Lerner, eds, *Traumatic Pasts: History, Psychiatry, and Trauma in the Modern Age, 1870–1930* (New York: Cambridge University Press, 2001).
64. Otis notes, however, that Wells's Moreau is more complex and sympathetic than Collins's Benjulia, reflecting Wells's ambivalent support for responsible uses of vivisection.
65. Grant Allen, *Physiological Aesthetics* (London: Henry King, 1877), 98.
66. Mark S. Micale and Roy Porter, eds, *Discovering the History of Psychiatry* (New York: Oxford University Press, 1994), 25.
67. See, for example, the immensely popular and widely respected work of Harvard Medical School Professor Herbert Benson, whose book *The Relaxation Response* (1975) promotes the cardiac benefits of meditation. Benson was instrumental in founding the Benson-Henry Institute for Mind Body Medicine at Massachusetts General Hospital. Similarly, New Haven surgeon Bernie Siegel has drawn attention to the benefits of meditation and group therapy for cancer patients in best-selling works like *Love, Medicine and Miracles: Lessons Learned about Self-Healing from a Surgeon's Experience with Exceptional Patients* (1986).

I. Catalysts

1

Howled out of the Country: Wilkie Collins and H.G. Wells Retry David Ferrier[1]

Laura Otis

On 17 November 1881, physiologist David Ferrier appeared in Bow Street Police Court, charged with "perform[ing] experiments, calculated to give pain to two monkeys, in violation of the restrictions imposed by the Vivisection Act."[2] Since 1873, Ferrier had been stimulating the brains of live animals, noting any correlation between the cortical area to which he applied electricity and the resulting movement of the animal. In 1876, Ferrier's maps of dogs' motor cortices had won him election to the Royal Society.[3] But according to the Home Secretary, Ferrier had broken the law. He had experimented on monkeys—displaying them at a major scientific conference—after they had recovered from anesthesia, but he had not applied for a Certificate B.

Ferrier's trial became a crucial test case of the 1876 Anti-vivisection Act, an experiment performed on a highly renowned scientist. Reflecting on Gustave Flaubert's trial in 1857, Dominick LaCapra writes, "trials . . . are . . . noteworthy instances of the social reception of cultural phenomena. They attest to the way these phenomena are read or interpreted in a decisive social institution and to the hermeneutic conventions operative therein. . . . A trial enables one to be somewhat more precise in investigating a '*mentalité*' or 'climate of opinion.'"[4] Like Flaubert, who had written *Madame Bovary*, the electrophysiologist Ferrier was believed to have committed an outrage against public morals. In Ferrier's short but widely reported trial, his work was narrated and interpreted, both by experts and those outside of physiology. Scientists and their critics wanted to learn once and for all whether public resistance could stop the spread of animal experiments.

As anti-vivisectionists, physiologists, and neurologists struggled for power during the 1880s and 90s, two novelists formed fictional worlds that recreated the debates of Ferrier's trial in new settings. Wilkie

Collins's *Heart and Science* (1883) and H.G. Wells's *The Island of Dr. Moreau* (1896) offer critiques of science far more complex and insightful than those of Ferrier's prosecutors. The novels differ in artistic quality, but both invoke Ferrier's research and the public's emotional response. Collins, an outspoken opponent of vivisection, created a propagandistic story that rises above its demonization of experimentalists only in a few reflective moments. Wells, a cautious supporter of animal experiments, offered a terrifying look at the implications of Ferrier's findings. When assessing neurophysiology and anti-vivisection as cultural movements, it is crucial to read these fictional representations in parallel with the public record. Unrestricted by legal codes and procedures, these literary trials challenge the values of science in ways that no actual trial ever could.

Dr. Ferrier's researches

In the 1820s, the French physiologist Pierre Flourens had removed large portions of animals' brains and discovered that the creatures could still move, although they were reduced to "complex machines."[5] Flourens had concluded that although the cortex must play a role in sensory perception and willful movement, it was not necessary for the production of motion. Mental faculties could not be assigned to any particular loci in the cortex, whose substance was everywhere alike and a small portion of which sufficed to carry out mental functions. Ferrier doubted Flourens's findings. The French scientist had worked mainly with birds and lower mammals, and for a neuroanatomist, he had been distressingly vague about which portions of the brains he had cut out. To Ferrier, a specialist in forensic medicine at King's College Hospital, these experiments, which had impressed physiologists five decades earlier, represented sloppy thinking.

The aim of Ferrier's studies, begun at the West Riding Lunatic Asylum in 1873, was to determine "whether the cerebrum, as a whole and in each and every part, contains within itself, in some mysterious manner inexplicable by experimental research, the possibilities of every variety of mental activity, or whether certain parts of the brain have determinate functions."[6] His wording makes his convictions clear. For Ferrier, knowledge of the brain meant knowledge of what its parts *did*. If those parts could not be structurally or functionally differentiated, scientists would never learn anything significant about the brain. Denial of localization suggested not just mysticism but insidious laziness, a desire not to know.

In the quest to prove that specific regions of the cortex controlled particular zones of the body, Ferrier was not the first to have thought of electrically stimulating the brains of live animals. In 1870, the Berlin scientists Gustav Fritsch and Eduard Hitzig had applied current to the exposed brains of dogs and produced "combined muscular contractions on the opposite side of the body."[7] With striking consistency, the stimulation of particular areas gave rise to specific movements. Emphasizing the reproducibility of their results, Fritsch and Hitzig claimed that they had found the "birthplace of the will to move."[8] Their findings showed not just that the substance of the brain was electrically excitable, which had long been denied, but that defined areas of the cortex drove particular muscle groups.

Believing that he could improve on Fritsch and Hitzig's work, Ferrier began investigating the brains of cats, dogs, rabbits, and guinea pigs. While he admired the Berlin electrophysiologists' studies, he believed that the scientists had not "sufficiently defined" the cortical areas they were stimulating or "the true character of the movements."[9] In his own experiments, Ferrier anesthetized dogs with chloroform or ether, then, using antiseptic surgical techniques, parted their skin, opened their skulls, and exposed their brains. Using an induction coil and blunted electrodes, he applied low levels of electricity to their brains' surfaces and observed their responses. Although he did not quantify the voltage or current levels, Ferrier tested the strength of his electrical stimulus by applying the electrodes to his own tongue, where they produced "a pungent, but quite bearable, sensation."[10] Because brain tissue does not have its own sensory receptors—as the scalp and skull do—Ferrier's experimental animals should have experienced pain only if their wounds were disturbed or became infected.

The nature of Ferrier's experiments made it difficult to study anesthetized animals, however. In his first article, he reported that all of his animals had been given chloroform or ether "from humane motives."[11] He learned quickly, though, that "[what] will cause intense and indefinite action in an animal non-narcotized, will excite only moderate and definite action in an animal sufficiently narcotized to abolish all sense of pain, and no effect at all on an animal deeply anaesthetized."[12] Once he had operated on an animal, he had a limited amount of time in which to conduct experiments, since the exposed cortex rapidly lost its excitability; in dogs, after about four days.[13] In mapping studies like Ferrier's, the experiments began when the brain was exposed and ended when the animal died. In his quest to learn how active, living brains controlled bodies, he began studying animals that had awakened from anesthesia.

When he described the animals' responses, Ferrier made it clear that he was eliciting not just movements but behavior. Because his critics maintained that his stimuli were reaching lower brain centers, giving rise to reflex actions, he needed to prove that he was producing the kinds of movements animals would *want* to make. "The movements . . . are purposive or expressional in character," he asserted, "and such as we should, from psychological analysis, attribute to ideation and volition if we saw them performed by others."[14] In dogs, electrization of area "1" (indicated on a map) caused the animal to advance its opposite hind leg "as in walking," and stimulation of area "9" provoked a "loud and distinct bark."[15] By stimulating specific spots on a cat's motor cortex, Ferrier was able to produce a "movement . . . such as the cat makes when striking a ball with its paw" and "spitting, and lashing the tail as if in rage."[16] As scientific writing, Ferrier's descriptions of his experiments convince readers of their validity because, drawing on their own experiences, they can easily envision the animals' responses. Like a good fiction writer, he brings his experimental world alive for the reader.

Although Ferrier urged physiologists to consider clinicians' observations, he emphasized that there was no substitute for animal experiments.[17] Only in a live, healthy animal could one watch the brain work in concert with the body. In the clinic, one could conduct autopsies and look for correlations between brain lesions and behavioral disruptions, but this purely observational approach had its limits. In the laboratory, Ferrier could cauterize tightly defined areas whose functions he had just tested, then watch for functional losses, but a naturally occurring tumor or stroke destroyed a much wider area, eliminating many functions at once. Although clinical studies—with human subjects—could show correlations between brain lesions in a given area and a corresponding loss of function, no number of cases could prove causality.

It is in the clinic, however, that Ferrier's work had its most immediate impact, and when he was brought to trial, neurologists became some of his staunchest defenders. Before the existence of detailed cortical maps, even when clinicians were certain that a patient harbored a growing tumor, they could not locate it accurately enough to open the skull and remove it. From the time that Ferrier began his studies, he had thought actively about how animal experiments and clinical observations could be combined to produce trustworthy knowledge of brain function. In 1874, he analyzed the functional losses and autopsy reports of epileptic, aphasic, and melancholic patients, comparing clinical correlations of structure with function to those he was seeing in the laboratory.[18] At the time of Ferrier's trial, the *British Medical Journal* printed a passionate editorial explaining the importance of his work for practicing physicians. "To the physiologist and physician," proclaimed the author, "[his studies] are, by mapping out the brain, as invaluable as a chart of an unknown region would be to an explorer."[19] The writer then offered several case histories in which Ferrier's correlations of structure with function had saved lives. In one, a two-and-a-half-year-old girl had suffered hemiplegia after a slight fall, but "the paralysis pointing, according to Dr. Ferrier's localization of functions of the brain, to injury or disease of a certain definite spot, the skull at this spot was, therefore, laid bare, and a depressed fracture was discovered."[20] Once the crushed bone and pus were removed, the girl fully recovered. By mapping the brains of dogs, cats, and monkeys, Ferrier was creating knowledge relevant to human beings, whose brains were not all that different.

No one knew better than Ferrier how powerfully his cortical maps supported Darwin's theory of natural selection, published a decade or two earlier. In his first article, he analyzed "correspondences" between areas representing the paws, eyes, and mouths of dogs, cats, and rabbits, reflecting that the differences he was seeing were "in accordance with the habits of the animal[s]."[21] In 1878, he asserted provocatively, "it will be my endeavor to show you that what is true of the monkey is strictly true also of man."[22] Besides horrifying anti-vivisectionists, who read in his accessible books how to cauterize dogs' brains, Ferrier's studies disturbed the public for the same reason that Darwin's did. They offered specific, irrefutable evidence for the similarity of people and animals. To anti-vivisectionists, this was no news; consciousness of this similarity was driving their campaign against animal experiments. But a second implication of Ferrier's work proved more unsettling. Since the time of Luigi Galvani, Europeans had known that electricity could make a frog's leg jerk, even if that leg were removed from the frog. But by applying electricity to the brains of live animals, Ferrier was making them *want* to move. He was demonstrating that electricity could replace a creature's will. Perhaps—his studies suggested—there was nothing sacred about the human will, not even human consciousness. A current barely perceptible to the tongue could generate what had been regarded as meaningful behavior. Ultimately, Ferrier's researches aroused the public for the same reason that audiences shuddered at *The Matrix*. Both suggested that the human brain could be exchanged for an electrical power source.[23]

The case against Ferrier

In the 1870s, as experimental physiology emerged as a field and social respect for science increased, individuals from a wide range of groups united to challenge scientists' power over animals. In their published condemnations of vivisection, the most common arguments were as follows: (1) Given the similarity of people to animals, why is it legal to experiment on dogs but not on human beings? (2) Lack of respect for animal life leads to lack of respect for human life. (3) Results found in animals (such as the susceptibility to poisons) are often inapplicable to people, whose physiology is significantly different. (4) We can learn more through clinical observation than through vivisection. (5) While scientists claim to be working for humankind, they are usually experimenting for selfish reasons: to win fame for themselves or to promote their own theories over those of rival investigators. (6) Scientists are arrogant in presuming themselves above moral and divine laws. When they raised these arguments, anti-vivisectionists invoked both the differences and the similarities between people and animals.

Victorians of both sexes and varying social levels joined the campaign against animal experiments. Queen Victoria opposed vivisection, as did Charles Dickens and Thomas Carlyle.[24] But the movement drew most heavily on the middle class, particularly on middle-class women, who identified with the trapped, tormented animals.[25] Certainly the defenders of vivisection knew that they were facing female opponents, but they may have misread the women's response. In an address to the Association for the Advancement of Medicine by Research in 1879, Vice President P.H. Pye-Smith explained, "[women] see in the helplessness of the dumb animals an appeal for pity, almost like that of childhood, and are justly indignant with the selfish cruelty so often exercised upon them."[26] He urged his fellow scientists to take heart, however, for "I have never found an intelligent woman who could not see the rights of the case when fairly explained to her."[27]

One intelligent woman remained unconvinced. In late 1875, frustrated by the R. S. P. C. A.'s unwillingness to take action against animal experiments, social activist Frances Power Cobbe founded the Society for the Protection of Animals Liable to Vivisection (the Victoria Street Society (VSS)).[28] Under Cobbe's leadership, the VSS contacted sympathizers in the House of Lords, who drafted an anti-vivisection bill closely following the society's suggestions.[29] According to the first draft of the bill, experiments on cats and dogs were forbidden; only scientists licensed by the Home Secretary could perform experiments that might cause animals pain; and experiments that allowed animals to recover from anesthesia could be conducted only if approved by designated referees such as the presidents of scientific societies.[30] When the bill was read and debate began in May 1876, British physiologists felt the power of the anti-vivisection movement for the first time. Shocked, Ferrier wrote to his friend Lyon Playfair, "if we were restricted to rabbits, many researches of vital importance to physiology and of the utmost value to man would be rendered impossible."[31] During the next months, intense lobbying occurred as the anti-vivisectionists sought to drive through their legislation and the scientists tried to amend it in their favor. On 15 August, the bill passed as The Cruelty to Animals Act of 1876, but not in its original form. According to the final bill, anyone who wanted to experiment on a live vertebrate had to apply to the Home Secretary. If his application was successful, he received a license valid for one year. Experiments on cats or dogs, or that allowed animals to recover from anesthesia, could be conducted only with special certification. Violators would be fined £50.[32]

If physiologists had lobbied the Lords on their own, the Act might have passed as originally drafted. In 1876, London and Cambridge physiologists founded a Physiological Society, of which Ferrier was one of the first members.[33] As an emerging profession, however, poorer in numbers than their opponents, the experimentalists needed help badly. Thanks to Ernest Hart, editor of the *British Medical Journal*, they received it from British doctors, who at Hart's urging wrote to their parliamentary representatives and local papers protesting that the Act was an insult to their professional judgment. Well connected and experienced with political lobbying, the medical men sensed a challenge to their authority, and clinicians who had never performed an experiment made sure that Ferrier could continue mapping dogs' brains.[34]

Against the arguments of the anti-vivisectionists, the physicians and physiologists brought their own to bear. First of all, nature was the "arch-vivisector."[35] Probably, an animal who died in a forest of natural causes suffered more than one who died in a lab to produce knowledge. Second, if people killed or exploited animals for food, clothing, transportation, even for recreation, why should they not be used to advance human knowledge? "More pain is caused by the whip of a London cab-driver in one day than is inflicted in any physiological laboratory in this country in the course of weeks," asserted a writer for *Nature*.[36] When appealing to the educated public, scientists relied especially on a third argument: real compassion meant looking beyond animal suffering to human suffering, overcoming one's emotions through force of reason. Fourthly, as Ferrier insisted, there could be no substitute for animal experiments, and there were things one simply could not learn from clinical observation. Invariably, anyone treating a disease performed experiments, so that if doctors and scientists didn't experiment on animals, they would end up vivisecting their patients. By nature, medicine involved experimentation. The question was, "are they to be popular experiments . . . blindly made upon hundreds or thousands of human beings, . . . or are they to be made upon a few animals in laboratories?"[37] Finally—cautiously—scientists invoked a fifth line of argument: people opposing animal experiments were ignorant and hostile to science. With lay audiences, scientists and doctors generally avoided this claim, but they used it freely when lobbying politicians.[38]

Between 1876 and 1881, 25–35 physiologists per year applied for licenses and certificates to perform vivisections.[39] Feeling unqualified to assess the proposals, the Home Secretary hired George Busk, a Fellow of the Royal College of Surgeons, as a consultant, and sometimes solicited the advice of James Paget, a strong supporter of animal experiments.[40] Anti-vivisectionists were outraged, but even under these circumstances approval was not guaranteed. In the five years between passage of the Act and Ferrier's trial, the Home Office rejected 27 applications, mainly those of young, unknown experimenters and those of investigators outside of the London–Cambridge scientific community.[41] While most unsuccessful applicants were young and/or Scottish (like Ferrier, who had earned his medical degree in Edinburgh in 1868), a few were prominent scientists, such as Gerald Francis Yeo, Professor of Physiology at King's College London.[42]

Feeling that the fox had been put in charge of the henhouse, the anti-vivisectionists developed new pressure tactics. First, they circulated petitions. In 1876, 805 of these, with 146,889 cumulative signatures, were sent to the House of Commons.[43] Second, they continued their drive to create legislation that would ban experiments on live animals. From 1876 until 1884, bills that would have forbidden all vivisections were introduced to parliament each year—and voted down.[44] With the medical and scientific lobbies at work, anti-vivisectionists could not muster the support they needed, and they lacked the widespread public sympathy they claimed to enjoy. When their legislation would not pass, they promoted their own candidates for parliament, trying to convince voters that an anti-vivisection M.P. would also support "the cause of the weak against the strong, . . . of right against might."[45] This didn't work either, although it might have if women had had the right to vote.

Since many leaders of the movement were well educated and had leisure time, they scrutinized the physiology journals in which the scientists published, looking to see whether their descriptions of work to their peers differed from their statements to the general public. In "Vivisection and Its Two-Faced Advocates" (1882), Frances Power Cobbe noted that in professional journals, the stated purpose of scientific studies was generally to show that "Professor A's theory has been disproved and that of Professor B (temporarily) established," whereas in popular review articles, the same studies were presented as heroic attempts to conquer diseases.[46] Once the Act went into effect, anti-vivisectionists compared the publications of English scientists with applications sent to the Home Office, which were a matter of public record.[47] Doubting the government's will to enforce the legislation, they tried to enforce it themselves.

At the International Medical Congress in London in August 1881, the scientists bonded and rallied. With over three thousand attendees, it was the largest medical meeting ever convened, and its debates received considerable press coverage.[48] In this highly visible forum, the German physiologist Friedrich Goltz challenged Ferrier's claim that specific brain regions controlled particular zones of the body, arguing that all cerebral tissue had the potential to carry out all of the brain's functions. To prove his theory of cerebral localization, Ferrier invited attendees to examine monkeys with functional disorders caused by brain lesions he had artificially induced.[49] When Cobbe and her colleagues at the VSS examined the Home Office records, however, they saw no sign that he had applied for a Certificate B, required of all experimenters who allowed animals to recover from anesthesia.[50] Citing the Act, the VSS brought formal charges.

Dr. Ferrier on trial

Ferrier's prosecution tested the Act's viability, since Ferrier was a widely respected scientist. "A trial is a force for intimidation," Dominick LaCapra has written, and Ferrier's case became a public confrontation in which the leading representatives of the scientific, medical, and anti-vivisection communities faced off.[51] Cobbe and her associates hoped to prove that no scientist, no matter how renowned, was above the law. Simultaneously, they wanted to publicize the details of Ferrier's researches, provoking public outrage. While maintaining Ferrier's innocence, the scientists and doctors aimed to show the futility of drafting laws that restricted science. The British Medical Association paid Ferrier's legal expenses, and its lawyers represented him in court.[52]

At the trial, "several ladies, among whom were Miss Frances Cobbe, interested in anti-vivisection, were accommodated with seats upon the bench." Police had to keep a "large crowd of medical students" outside of the courtroom because of their "howling and cheering."[53] Representing the VSS, the prosecuting attorney Mr. Waddy presented his case as follows:

> [Ferrier] had failed to take out the necessary certificate required by the Act. . . . Certain animals were operated upon six months ago, and instead of being killed before the ceasing of the anesthetic power under which they were placed, they were kept alive for the purpose of other experiments being tried upon them. . . . Now, experiments should be made to ascertain an immediate fact, and the animal be killed before the effect of the anesthetic had ceased. . . . In this case, however, the experiments began when the monkey woke up, and observations were made from time to time.

He accused Ferrier of "participation in a continuous experiment."[54] No one with an even rudimentary grasp of physiological experimentation could have made such an argument. Physiological experiments are "continuous" by nature, documenting changes in the way an animal lives after its body has been altered. In mapping studies, the anesthetic is used so that the animal feels no pain when its brain is exposed. In Ferrier's investigations of voluntary movements, it was nearly impossible to obtain "immediate facts" while the animals slept; reliable data could be gathered only after the animals had awakened. In this face-off between scientists and the defenders of animals, the plaintiff's case suggests how poorly the activists may have understood what they were attacking.

A second tactic by the prosecution may indicate a wider gulf. To specify how Ferrier had used surgically altered monkeys at the International Medical Congress, Waddy summoned two prominent scientists, questioning them about the differences of opinion between Goltz and Ferrier. Cambridge physiologist Michael Foster, President of the Physiological Section, recalled:

> Professor Goltz opened the discussion, holding that the surface of the brain acts as a whole, and Professor Ferrier held that any act only called into action a special part of the brain. Facts in support were brought forward on both sides. Dogs and monkeys were both referred to as having been operated on. . . . It was thought desirable that the many foreign physiologists present should see the dog and monkeys, . . . and I . . . arranged the *séance* at King's College Hospital.[55]

Waddy's emphasis on the Ferrier–Goltz debates suggests that he was after more than the details of the demonstration. Cobbe had expressed her disgust with scientists who declared publicly that they were trying to cure diseases, yet published articles disproving Professor A's theory and temporarily establishing Professor B's. What bothered many anti-vivisectionists was that these debates occurred at all. Supposedly, scientific knowledge was certain, built on established facts. If scientists could not agree, if "facts in support" could be offered on each side, if scientific knowledge involved *interpretation* and could be declared worthless in light of future findings, how could scientists justify killing monkeys and dogs? What were they producing, in comparison to the value of an animal's life?

Waddy's choice to probe the Ferrier–Goltz debates can thus be read in two senses. Given his protest against "a continuous experiment," it may indicate a misunderstanding of what science is. Perhaps he hoped to imply that where there was disagreement, where interpretation was all-important, the competence of the scientists was in question. On the other hand, he may have understood scientific inquiry all too well, attacking it at what he perceived to be its weakest point. In a system of knowledge production that presented itself to the public as certain and factual, disagreement was a serious problem.

Once it emerged that Yeo, not Ferrier, had performed the surgery on the monkey's brain, the prosecution's case collapsed, for unlike Ferrier, Yeo had obtained the necessary certificate. Judge Ingham dismissed the case, and Ferrier did not have to pay the £50 fine.

Wilkie Collins retries Ferrier: *Heart and Science* (1883)

A personal friend of Cobbe and member of the VSS, Wilkie Collins wrote *Heart and Science* not long after Ferrier's trial. In his preface, he thanks Cobbe for her "assistance" in preparing the book, although he does not specify how she helped.[56] More significantly, he quotes the conclusion of Ferrier's *The Localization of Cerebral Disease* (1878), in which the physiologist admits, "we cannot even be sure, whether many of the changes discovered are the cause or the result of the disease, or whether the two are the conjoint results of a common cause."[57] Since Collins offers the reader neither page reference nor context, it is uncertain whether he read the book or understood Ferrier's aims. Probably Cobbe assisted by giving him the quotation, which he used again in the main text of the novel. His sarcastic response to it ("Plenty of elbow room here for the spirit of discovery") suggests he believed that if scientists doubted the meaning of their results, they had no right to be killing dogs.

Almost no one has ever thought that *Heart and Science* is a good novel. Valerie Pedlar, who has also pointed to Ferrier's work as a source, has pronounced it "an unashamed piece of polemic cast in fictional form."[58] As the preface indicates, it was written to prove a point rather than to tell a story.

In the novel, the talented physician Ovid Vere is about to collapse from overwork and is planning a trip abroad when he sees a beautiful young woman on a London street. He soon learns that the woman, Carmina, is his own cousin, the child of his mother's brother and an Italian actress, who married despite the English family's disapproval. Almost immediately, Ovid and Carmina fall in love. Ovid's mother, Mrs. Gallilee, is monstrously jealous of her younger sister, Lady Northlake, whose marriage to a Lord preceded her own to Ovid's father. When Ovid's father died, she married weak-willed Mr. Gallilee, with whom she now has two daughters, the affected model child Maria and the lovable basket-case, Zo. A scientific enthusiast, Mrs. Gallilee has gotten the family into financial difficulties in her efforts to organize scientific *soirées* and outshine her sister. When she learns that Carmina is to inherit her recently deceased brother's fortune, on which she had been counting, she proves diabolical in her attempts to show that Carmina is not the legitimate heir. Meanwhile, the sadistic physiologist Dr. Benjulia, an acquaintance of Ovid and his mother, has been conducting experiments on monkeys and dogs in an effort to understand brain disease. While Ovid is in Canada convalescing, he attends Carmina, who has been driven nearly mad by Mrs. Gallilee's abusive treatment, and almost lets her die in his eagerness to watch her disease progress. In Montreal, Ovid is given the manuscript of an unknown physician who has found a cure for brain disease through clinical observation alone. When Ovid returns and publishes it, and Benjulia learns that he has lost priority despite years of vivisections, he burns his lab and commits suicide. Ovid marries Carmina, and with her son's support, Mrs. Gallilee continues to give scientific *soirées*.

From the title, it would appear that Collins is creating a dichotomy, his characters falling into either the "heart" or the "science" camps. In the novel, animals are often wiser, more perceptive, and more compassionate than people, and individuals' responses to them serve as moral measuring sticks of their sympathy, altruism, and generosity. Carmina faints when she sees a stray dog run over by a cab. Adorable Zo, who is introduced at the zoo and whose name suggests animal impulses, inspires warm feelings in everyone but Benjulia, who likes to experiment with her spine. "I don't like dogs," says Benjulia. "They bark" (1: 184).

Over the course of the novel, however, some of Collins's characters develop and grow. Mrs. Minerva, a skinny governess in love with Ovid, overcomes her jealousy of Carmina and helps her to escape Mrs. Gallilee's clutches. Even Benjulia experiences confused feelings of affection for Zo, thinking before he ends his life that he wishes he could tickle her once more.[59] Collins's oppressive thesis that love inspires and vivisection degrades helps his character development, for at least some of his creations change their ways.

The wicked Benjulia does not. Because of Collins's tactic of tying bad science to a personality, his fictional vivisector cannot change, even though Collins hints that he has the potential to become a better man.[60] In his appearance and work habits, Benjulia seems more like a melodramatic villain than any London physiologist. The vivisector stands six foot six and is as thin as a skeleton, with "the hair of an American Indian" and a complexion of "true gipsy brown" (1: 108). In his looks, he is a "gipsy beggar" in the same sense as Brontë's Heathcliff, an outsider of unidentifiable origins.[61] Benjulia lives and conducts his experiments in "a hideous square building . . . in the middle of a barren field" in a London suburb (1: 177). When Ovid visits him, he finds that "there was something unnatural in the solitude of the place" (1: 178). Working in isolation, he makes sure that his few servants never enter his laboratory, from which Ovid hears "a moaning cry" (1: 178). Disliking women, possessing no family or friends, Benjulia searches obsessively for the cause of an unspecified brain disease, torturing countless animals in the process.

Significantly, Collins never takes readers into the forbidden lab, possibly because he knew he was not competent to describe it. Instead, he invites them to imagine Benjulia's vivisections by observing the way that he treats people, which certainly suggests horrors.[62] As a character, Benjulia fulfills the anti-vivisectionists' prediction that people who conduct or even witness animal experiments will perform cruel experiments on human beings. His manipulations of Zo's spine recall Ferrier's brain-mapping studies:

> He put two of his soft big finger tips on her spine, just below the back of her neck, and pressed on the place. Zo started and wriggled under his touch. He observed her with as serious an interest as if he had been conducting a medical experiment. "That's how you make our dog kick with his leg," said Zo.
>
> (1: 111)

Of course, Ferrier stimulated exposed brains, and Benjulia is manipulating an intact spinal cord, but the correlation between a nervous stimulus and a specific movement suggests that Collins did read Ferrier's work. In *The Functions of the Brain* (1876), the physiologist had reported that when he stimulated the first convolution of a dog's left hemisphere, "the opposite hind leg is advanced as in walking."[63]

Through Benjulia, Collins promotes the anti-vivisectionists' claim that scientists are selfish, a notion easier to convey through fictional associations than through public statements. Interrogated by his brother in the novel's central scene, Benjulia finally admits, "I am working for my own satisfaction—for my own pride—for my own unutterable pleasure in beating other men—for the fame that will keep my name living hundreds of years hence" (2: 11). Consequently, Benjulia feels no sense of solidarity with other scientists, viewing them only as rivals and threats. He subscribes to "every medical publication in England" but does not read the journals, merely scanning them for any indication that another investigator has encroached on his turf (2: 269). When he learns that an unknown physician has preceded him in elucidating brain disease, he tells Ovid, "you have taken something from me, which was dearer than life" (2: 274). In the novel's climactic scene, a horrified servant discovers the real nature of Benjulia's research as "black figures of small swiftly-running creatures" and a limping dog escape from his laboratory (2: 280). In his death, Benjulia portrays the hypocrisy of vivisectors, for he takes opium before burning his lab, unwilling to face the pain he has been inflicting on animals.

Science in Collins's novel does not end with Benjulia. It has other representatives, none worse, and some better. Through the figure of Mrs. Gallilee, Collins goes after the premise that scientists and their admirers are more intelligent than their critics. Her infatuation with science springs not from intellectual curiosity but sibling rivalry:

> From the horrid day when [her sister] became Lady Northlake, [Mrs. Gallilee] became a serious woman. All her earthly interests centered now in the cultivation of her intellect. She started on that glorious career, which associated her with the march of science. In only a year afterwards . . . she was familiar with zoophyte fossils, and had succeeded in dissecting the nervous system of a bee.
>
> (1: 62)

In her devotion to science, she shows no more discrimination than a religious fanatic or an obsessed lover. Her judgment is poor, and she does not seem particularly bright. On several occasions, Collins suggests that charlatans are using her, presenting their work as science so as to obtain her patronage. When Carmina picks up a book with "beautiful binding," Mrs. Gallilee tells her, "You have taken up the 'Curiosities of Coprolites.' That book is one of my distinctions—a presentation copy from the author." Apologizing for her ignorance, Carmina asks what coprolites are, and Mrs. Gallilee tells her patronizingly, "Coprolites are the indigestions of extinct reptiles" (1: 142). Much of science is dinosaur dung, but Mrs. Gallilee cannot distinguish the reptile from its droppings.

By associating scientific attitudes with these two characters, Collins avoids condemning science as a whole. He attacks only a certain kind of science, in which the desire to know is divorced from basic human emotions. In contrast to Benjulia and Mrs. Gallilee, Collins offers the unknown American doctor who discovered the secrets of brain disease by observing suffering patients. Ovid, who attends the dying physician, writes to Carmina, "what his malady was, you will not ask to know." Readers can guess at its nature, though, when they learn that his wife had committed "every worst offence of which a bad woman can be guilty" (1: 234).[64] Exhibiting dog-like loyalty, the martyred physician had refused to leave his wife and had fallen into poverty and illness. Still, he had managed to write a book of the highest value to neurologists, declaring in his preface, "I am innocent, in my professional capacity, of ever having perpetrated the useless and detestable cruelties which go by the name of Vivisection" (2: 250). Through this description, Collins implies that the greatest contributions to science come from compassionate individuals whose own suffering lets them identify with laboratory animals.

In a society in which science commands so much respect, Collins warns, the danger lies in granting it too much credence, in reading science uncritically. In a didactic passage, his narrator admonishes the "learned lady":

> See the lively modern parasites that infest science, eager to invite your attention to their little crawling selves. Follow scientific inquiry, rushing into print to proclaim its own importance, and to declare any human being, who ventures to doubt or differ, a fanatic or a fool. . . . Submit to lectures and addresses by dozens which, if they prove nothing else, prove that what was scientific knowledge some years since, is scientific ignorance now—and that what is scientific knowledge now, may be scientific ignorance in some years more.
>
> (2: 206)

Like Ferrier's prosecuting attorney, Collins' narrator points out the fallibility and evanescence of scientific knowledge. If its critics are dismissed as ignorant, he asks, who will be left to challenge its abuses? Clumsy as it is, his novel reiterates the central questions of Ferrier's trial in a more direct, compelling form. Who is fit to judge the quality of scientific work? Who besides scientists can police science?

H.G. Wells retries Ferrier: *The Island of Dr. Moreau* (1896)

Between 1883 and 1896, British scientists gained power. In March 1882, stimulated by Ferrier's trial to organize and defend their interests, leading physiologists and physicians formed the Association for the Advancement of Medicine by Research.[65] When challenged by anti-vivisectionists, they avoided confrontation, focusing their efforts on political lobbying. As a result, three decades of activism failed to achieve a ban on animal experiments, or even to slow the growth of experimental physiology. Believing in the moral value of their cause, however, the anti-vivisectionists fought on, publishing pamphlets on the work of particular offenders. Two of these were *Professor Ferrier's Experiments on Monkeys' Brains* (1885) and *Ferrieristic Brain Surgery: A Candid Condemnation* (1887).[66]

The first pamphlet described the suffering of the monkeys Ferrier had exhibited at the 1881 meeting. "The truth is out at last—Professor Ferrier has explained!" it began, offering Wells a good title for his central chapter.[67] The author reported that Ferrier and Yeo had cauterized the brains of monkeys with a "red hot wire," inducing blindness, then watched as the animals ran around in terror, crashing into furniture. "And so for two hours did these English gentlemen (!) torture this poor blinded creature," ranted the writer, "suffering as it was from the effects of a terrible operation, and in mortal dread of its tormentors! . . . Professor Ferrier was acquitted at Bow Street, [but] . . . there is a larger tribunal before which both he and his colleague must appear."[68] In comparison to this tirade, Wells's depiction of Moreau is ambivalent, but like the pamphlet, it encourages readers to question the motives, methods, and value of physiological research.

Unlike Collins, Wells knew a great deal about experimental physiology, having studied biology with Thomas Henry Huxley at the Normal School of Science in 1884. In his *Experiment in Autobiography*, Wells wrote that he loved Huxley's course because it was "a vivid, sustained attempt to see life clearly and to see it whole, to see into it, to see its inter-connections, to find out . . . what it was, where it came from, what it was doing and where it was going."[69] Wells supported responsibly conducted animal experiments, and like Huxley, he believed in Darwin's theory of natural selection.[70] By creating Moreau, whose experiments were anything but responsible, Wells was not condemning science in general, nor was he referring explicitly to Ferrier's research. His novel is less about the cruelty of scientists than the cruelty of nature.[71] In its implications, however, *The Island of Dr. Moreau* comes much closer to Ferrier's findings than to Collins's work that specifically invokes them. Like Ferrier's studies, Wells's novel suggests that there is nothing holy about human consciousness, and that people and animals cannot be separated by any identifiable boundary.

From the beginning of Wells's work, people behave like animals. After the narrator Prendick, his companion Helmar, and a sailor drift for days in a lifeboat, they draw lots to see which of the three will be eaten by the other two. When the sailor rejects his fate and falls overboard in a fight with Helmar, the delirious Prendick laughs. Soon afterward, the drunken captain throws him off of the ship that has rescued him, a freighter carrying animals to a small island northwest of Peru. Seeing the captain's "dull and solemn eyes," Prendick thinks disgustedly, "the brute was drunk."[72] Throughout the novel, alcohol recurs as a portal between the bestial and the human, emphasizing the fluidity of these categories.[73]

On this remote island, the vivisector Moreau is trying to turn animals into humans, as he has been attempting for two decades. As many critics have noticed, Moreau's experiments seem like an "allegory of the evolutionary process."[74] By depicting a scientist who causes excruciatingly painful transformations, Wells emphasizes the elements of "chance, waste, and pain" in the evolutionary process, in which change occurs through random variations and animals with harmful variations die painful deaths.[75] His novel dramatizes the *British Medical Journal*'s claim that the arch-vivisector is nature.

As in Collins's story, science does not begin or end with one vivisector. Wells's tale offers three scientists: Moreau; his assistant, Montgomery; and the narrator, Prendick.[76] In his scientific training, Prendick initially resembles Wells, having "spent some years at the Royal College of Science, and . . . done some researches in biology under Huxley" (18). The similarity ends there, however.[77] Unlike Wells, who came from a poor family and was able to study only on scholarship, Prendick tells readers, "I had taken to natural history as a relief from the dullness of my comfortable independence" (7). Not surprisingly, Prendick lacks the skills necessary to a good naturalist, confessing to readers, "my eye has no training in details, and unhappily I cannot sketch," "a certain lack of practical sense . . . has always been my bane," and "I am an extremely unhandy man" (53, 83, 84). While upper-class readers may have identified with Prendick, one hears Wells's voice more in Moreau, who tells the amateur, "I have seen more of the ways of this world's Maker than you—for I have sought his laws, in *my* way, all my life, while you, I understand, have been collecting butterflies" (48).

Montgomery fares no better as a scientist, although for different reasons. Prendick realizes that "he had evidently been a very ordinary medical student," learning his biology at University College (7). He is serving Moreau on the island instead of patients in London "simply because—eleven years ago—[he] lost [his] head for ten minutes on a foggy night" (12). Like the Beast-Folk, Montgomery has been remolded by Moreau, and he defends and sympathizes with them more than any of the other human characters. Seeing the pain that Moreau is causing, Montgomery drinks because he hates what he is doing but can see no way out of the situation in which he has landed. In his drunkenness, he resembles Moreau's other creations.

In comparison to these two, Moreau is a competent, thinking scientist. He has withdrawn from the British medical community not because his experiments were poorly conceived, but because anti-vivisectionists "howled him out." Prendick recalls that Moreau was once:

> A prominent and masterful physiologist, well known in scientific circles for his extraordinary imagination and his brutal directness in discussion. . . . Then suddenly his career was closed. He had to leave England. A journalist obtained access to his laboratory in the capacity of a laboratory assistant, with the deliberate intention of making sensational exposures; and by the help of a shocking accident (if it was an accident) his gruesome pamphlet became notorious. On the day of its publication, a wretched dog, flayed and otherwise mutilated, escaped from Moreau's house. It was in the silly season, and a prominent editor, a cousin of the temporary laboratory assistant, appealed to the conscience of the nation. It was not the first time that conscience has turned against the methods of research. The doctor was simply howled out of the country.
>
> (21–2)

Like Ferrier, Moreau has been experimenting on dogs, and Wells's description of his treatment suggests that he had the neurophysiologist in mind. Moreau's work was demonized in "a little buff-colored pamphlet, to read which made one shiver and creep" (21). In Wells's representation, however, readers cannot tell whether Moreau's London experiments were truly sadistic. It was the anti-vivisectionists who acted like cruel, baying hounds.

Although Moreau is more complex than Collins's Benjulia, Wells describes his scientist's selfishness in similar terms. Prendick tells us that, "[Moreau] was unmarried, and had indeed nothing but his own interest to consider" (22).[78] Moreau never explicitly says why he is trying to turn animals into human beings, but since he has withdrawn from the scientific community, it is unlikely that he will tell anyone his results or is working to relieve human suffering. When Prendick asks him, "Where is your justification for inflicting all this pain?" Moreau dodges the question, telling Prendick that he doesn't understand what pain is. "I went on with this research just the way it led me," he confesses to Prendick. "I asked a question, devised some method of obtaining an answer, and got—a fresh question" (47–8). Apparently, he is motivated by sheer curiosity.[79]

Prendick senses what Moreau is doing long before Moreau "explains" himself in the central chapter, but significantly, he gets it backwards. In the first half of the story, Prendick is repulsed by the Beast-Folk, but simultaneously, he feels an affinity toward them. When he encounters three of them jabbering in the forest, he tells readers, "I perceived clearly for the first time what it was that had offended me, what had given me the two inconsistent and conflicting impressions of utter strangeness and yet of the strangest familiarity." Recognizing their humanity, he recalls, "the most horrible questionings came rushing into my mind" (27). When Prendick does form a hypothesis, he opts for the less threatening possibility that Moreau is turning people into beasts. "They were men," he protests, "men like yourselves, whom you have infected with some bestial taint—men whom you have enslaved, and whom you still fear" (43). Unable to conceive that animals can be turned into people, Prendick imagines a scenario less frightening and—given the behavior he has been seeing—more realistic. When Moreau tells him the truth, he refuses to believe it.

To convince him, Moreau must create a scientific narrative, explaining his work on analogy with plastic surgery. Sounding a lot like Darwin's anthropomorphized natural selection, he tells Prendick, "these creatures you have seen are animals carven and wrought into new shapes. To that, to the study of the plasticity of living forms, my life has been devoted" (46). In the mid-1890s, with the field of plastic surgery coming into its own, Moreau's references to grafting and antisepsis would have seemed plausible.[80] If Moreau's work is successful, humanity is only skin deep.

As Moreau has found, however, sculpting the animals' forms is not enough. To complete the metamorphosis, he has also had to change their mental structure. It is here that Ferrier's discoveries become most relevant to Moreau's research, although Wells's physiologist is practicing hypnosis, not brain surgery. "The possibility of vivisection does not stop at a mere physical metamorphosis," claims Moreau:

> A pig may be educated. The mental structure is even less determinate than the bodily. In our growing science of hypnotism we find the promise of a possibility of superseding old inherent instinct by new suggestions, grafting upon or replacing the inherited fixed ideas. Very much indeed of what we call moral education . . . is such an artificial modification and perversion of instinct.
>
> (47)

To keep his newly forged humans from reverting to their animal habits, he has inculcated laws, which the animals recite to maintain their new, unnatural identities. As a scientific observer, however, he admits that, "as soon as my hand is taken from them, the beast begins to creep back" (51). If there is a difference between people and animals, Wells's novel suggests, it is not speech, reason, or intellect. In his descriptions of people and animals, he offers the reader no fixed point of reference. There is no model human behavior, no outside-of-the island, that would allow an interpreter to define qualities as typically animal or human.

In the second half of the novel, as Moreau's law disintegrates and the Beast-Folk turn on their creator and his hapless assistant, Prendick's descriptions make it increasingly clear that he is thinking of another island.[81] Sympathizing with the Beast-Folk, who still disgust him, Prendick tells us, "I would see one of the clumsy bovine creatures who worked the launch treading heavily through the undergrowth, and find myself asking, trying hard to recall, how he differed from some really human yokel trudging home from his mechanical labors" (55). The most terrifying moments of Wells's novel come not when Prendick is fleeing the Beast-Folk but when he returns home and still feels that he is among vivisected beasts. In London, he notices that

> Prowling women would mew after me, furtive craving men glance jealously at me; weary, pale workers go coughing by me with tired eyes and eager paces, like wounded deer dripping blood . . . And even it seemed that I too was not a reasonable creature, but only an animal tormented with some strange disorder in its brain.
>
> (87)

In this final description, Prendick presents himself and all of his fellow human beings as vivisected animals, controlled not by their own wills but by the stimuli of the perverse vivisector, nature. Prendick responds by withdrawing, abandoning biology in favor of chemistry, astronomy, and "wise books" (87). Seeking comfort in a higher law, he is more like the Beast-Folk than ever.

Conclusion

I have argued that Collins and Wells retry Ferrier by creating fiction that recalls and challenges his science. As art, however, literature never emerges purely from a desire to prove a point. In the wake of Ferrier's trial, Collins tried to move readers emotionally, seeking allies in his fight to end animal experiments. By associating arguments and attitudes with characters he created, he gained a psychological advantage he could never have wielded in a pamphlet. Still, *Heart and Science* is more than propaganda. As in Ferrier's trial, the reader hears multiple voices offering radically different views of animal experiments. Unlike Ferrier's prosecutor, Collins's narrator asks readers straight out whether a short-lived scientific truth is worth an animal's life.

Wells's story also demands an emotional response, but it would do him a great injustice to claim that he wrote *The Island of Dr. Moreau* to condemn vivisection. Like Collins's novel, Wells's work depicts the potential ravages of science unchecked, but it also caricatures obsessed anti-vivisectionists. *The Island of Dr. Moreau* is better read as a literary response to Huxley's questions of what life is, where it comes from, what it is doing and where it is going. The buff-colored pamphlet, the "howling" anti-vivisectionists, the attempts to control behavior, and the recitation of the law all recall Ferrier's trial, but Wells shows no desire to manipulate readers into supporting or opposing animal experiments. Vivisection in Wells's novel serves more as a metaphor for the human condition, in which people's animal nature is suppressed through excruciating pain but keeps emerging despite inculcated laws. *The Island of Dr. Moreau* retries Ferrier by questioning the role of science in a world in which people live as vivisected animals. Through dialogues among characters with varying degrees of perceptiveness, Wells experiments with science, encouraging readers to rethink its purpose in a cruel universe.

Notes

1. I am grateful to the John D. and Catherine T. MacArthur Foundation for the grant that allowed me to conduct the research presented in this essay, and to the Max Planck Institute for the History of Science in Berlin, which hosted me while I conducted it. I thank my colleagues at the MPI, particularly Skuli Sigurdson, for their comments on the manuscript. I am indebted to my colleague, Paul Lennard, and to our Emory University students in "The Roots of Modern Neuroscience" for their comments on Ferrier's work.
2. "The Charge Against Professor Ferrier," *The London Times* (18 November 1881), 10.
3. Robert M. Young, *Mind, Brain and Adaptation in the Nineteenth Century: Cerebral Localization and Its Biological Context from Gall to Ferrier* (New York: Oxford University Press, 1990), 239.
4. Dominick LaCapra, *Madame Bovary on Trial* (Ithaca: Cornell University Press, 1982), 15–6.
5. All information on Flourens's experiments in this paragraph is from David Ferrier, *The Functions of the Brain* (1876; London: Dawson's, 1966), 125;

Edwin Clarke and L.S. Jacyna, *Nineteenth-Century Origins of Neuroscientific Concepts* (Berkeley: University of California Press, 1987), 244–66; and Young, *Mind, Brain and Adaptation*, 226.
6. Ferrier, *The Functions of the Brain*, 124–5.
7. G. Fritsch and E. Hitzig, "Über die elektrische Erregbarkeit des Grosshirns," *Archiv für Anatomie, Physiologie und wissenschaftliche Medicin* 37 (1870), 300–32, 311. See also Anne Harrington, *Medicine, Mind, and the Double Brain: A Study in Nineteenth-Century Thought* (Princeton: Princeton University Press, 1989), 70–1; and Young, *Mind, Brain, and Adaptation*, 224–33. Translations of Fritsch and Hitzig's article are my own.
8. Fritsch and Hitzig, "Über die elektrische Erregbarkeit," 326.
9. Ferrier, *The Functions of the Brain*, 131.
10. Ferrier, *The Functions of the Brain*, 129.
11. David Ferrier, "Experimental Researches in Cerebral Physiology and Pathology," *West Riding Lunatic Asylum Medical Reports* 3 (1873), 30–96, 33.
12. Ferrier, *The Functions of the Brain*, 129–30.
13. David Ferrier, *The Localization of Cerebral Disease* (London: Smith, Elder, 1878), 29.
14. Ferrier, "Experimental Researches," 72–3.
15. Ferrier, *The Functions of the Brain*, 149–50.
16. Ferrier, *The Functions of the Brain*, 154–5.
17. Ferrier, *The Functions of the Brain*, xiv.
18. David Ferrier, "Pathological Illustrations of Brain Function," *West Riding Lunatic Asylum Medical Reports* 4 (1874), 30–62.
19. "Dr. Ferrier's Researches," *The London Times* (19 November 1881), 12.
20. "Dr. Ferrier's Researches," 12.
21. Ferrier, "Experimental Researches," 60–1.
22. Ferrier, *The Localization of Cerebral Disease*, 31.
23. In Ferrier's studies, electricity replaces impulses that would normally originate in the motor cortex. In *The Matrix*, this situation is reversed, and the brains of enslaved human beings power alien technology.
24. Coral Lansbury, "Gynaecology, Pornography, and the Antivivisection Movement," *Victorian Studies* 28 (1985), 413–37, 414.
25. Lansbury argues that women campaigned passionately against vivisection because they saw themselves in the dogs strapped to scientists' tables. Victorian gynecologists often restrained women, placing their feet in stirrups so that they were immobilized in postures like those of vivisected animals. In Victorian pornography, whip-bearing "riding masters" raped women immobilized by saddles and stirrups. Lansbury, "Gynaecology," 418–22. Mary Ann Elston reports that science was depicted as rape in anti-vivisection literature, and women "were explicitly invited to identify themselves with the animals." She qualifies, however, that anti-vivisection was "never exclusively a woman's cause." Mary Ann Elston, "Women and Anti-vivisection in Victorian England, 1870–1900," in *Vivisection in Historical Perspective*, ed. Nicolaas A. Rupke (New York: Routledge, 1990), 263, 279.
26. P.H. Pye-Smith, "Address," *Nature* (28 August 1879), 408–13, 411. I have taken the liberty of substituting "women" for the word "they." In this paragraph, Pye-Smith is referring to "another class of persons, less ignorant and less stupid," who oppose vivisection because they allow their sentiments to

overpower their reason. His wording, especially the reference to children, suggests that he means middle-class women.

27. Pye-Smith, "Address," 411.
28. Émilie Dardenne, "Frances Power Cobbe (1822–1904) et la naissance du mouvement contre l'éxperimentation animale," *Cahiers victoriens et edouardiens* 59 (2004), 171–80, 171–2.
29. Richard D. French, *Anti-vivisection and Medical Science in Victorian Society* (Princeton: Princeton University Press, 1975), 113–4.
30. French, *Anti-vivisection*, 115.
31. French, *Anti-vivisection*, 120.
32. French, *Anti-vivisection*, 143.
33. Edwin Clarke, "David Ferrier," *Dictionary of Scientific Biography*, ed. Charles Coulston Gillispie, 14 vols (New York: Charles Scribner's Sons, 1970–80), 4: 593–5, 593.
34. French, *Anti-vivisection*, 130–2, 157.
35. "Dr. Ferrier's Researches," 12.
36. "Vivisection and Medicine," *Nature* (11 August 1881), 329–32, 331.
37. "Vivisection and Medicine," 331.
38. French, *Anti-vivisection*, 217.
39. French, *Anti-vivisection*, 184–8.
40. French, *Anti-vivisection*, 179–80.
41. French, *Anti-vivisection*, 188–9.
42. Clarke, "David Ferrier," 4: 593; French, *Anti-vivisection*, 186, 189. Although trained in Scotland, Ferrier was by no means an outsider in the London scientific establishment, having held a chair of forensic medicine at King's College Hospital since 1872.
43. French, *Anti-vivisection*, 168.
44. French, *Anti-vivisection*, 164.
45. French, *Anti-vivisection*, 166.
46. Frances Power Cobbe, "Vivisection and Its Two-Faced Advocates," *The Contemporary Review* 41 (1882), 610–26, 611.
47. French, *Anti-vivisection*, 175.
48. Nicolaas Rupke calls this congress a "clarion call to organized self-defense" by European scientists and physicians who saw their social and professional prestige threatened by anti-vivisectionists. Nicolaas A. Rupke, "Pro-vivisection in England in the Early 1880s: Arguments and Motives," in *Vivisection in Historical Perspective*, ed. Nicolaas A. Rupke (New York: Routledge, 1990), 190.
49. To show that particular regions of the cortex did not control motor functions, Goltz exhibited decorticate dogs that could move in response to stimuli. However, Ferrier's demonstration of monkeys that moved when particular regions were stimulated proved more convincing, and most scientists who witnessed the demonstrations left believing that motor functions were driven by particular areas of the cortex. Harrington, *Medicine, Mind, and the Double Brain*, 117.
50. French, *Anti-vivisection*, 200–1.
51. LaCapra, *Madame Bovary on Trial*, 16.
52. French, *Anti-vivisection*, 202.
53. "The Charge against Professor Ferrier," 10.
54. "Dr. Ferrier Accused," *The New York Times* (11 December 1881), 5.

55. "The Charge against Professor Ferrier," 10.
56. Coral Lansbury reports that Cobbe "provided [Collins] with a large selection of antivivisection books and pamphlets." Lansbury, "Gynaecology," 427.
57. Wilkie Collins, *Heart and Science: A Story of the Present Time,* 2 vols (Leipzig: Bernard Tauchnitz, 1883), 1: 11. Subsequent citations of this work will be indicated by parenthetical references in the text.
58. Valerie Pedlar, "Experimentation or Exploitation? The Investigations of David Ferrier, Dr. Benjulia, and Dr. Seward," *Interdisciplinary Science Reviews* 28 (2003), 169–74, 172.
59. Coral Lansbury reads Benjulia's urges in a different way, making a convincing case that he is a pedophile and sexual sadist. In Victorian pornography, "tickling" was a euphemism for sex. Lansbury, "Gynaecology," 427–31.
60. Valerie Pedlar observes that in *Heart and Science,* "criticism of vivisection is made through character." Pedlar, "Experimentation or Exploitation?" 169. Coral Lansbury believes that "the real terror in *Heart and Science* is not in the locked and secret laboratory, but in the mind of Benjulia." Lansbury, "Gynaecology," 431.
61. The name "Benjulia" suggests that Collins's scientist may be Jewish, although there is no textual evidence to support this idea.
62. Coral Lansbury observes that "Collins is careful never to describe an experiment, deflecting the existing emotional response from the subject of the vivisected animal to the figure of the vivisector." Lansbury, "Gynaecology," 431.
63. Ferrier, *The Functions of the Brain,* 149.
64. Probably the doctor is dying of tertiary syphilis.
65. French, *Anti-vivisection,* 204.
66. French, *Anti-vivisection,* 281.
67. *Professor Ferrier's Experiments on Monkeys' Brains* (London: Victoria Street Society, 1885), 1.
68. *Professor Ferrier's Experiments on Monkeys' Brains,* 3, 7.
69. H.G. Wells, *Experiment in Autobiography* (New York: MacMillan, 1934), 169.
70. On Wells's attitude toward animal experimentation and evolutionary theory, see Roslynn D. Haynes, "Wells's Debt to Huxley and the Myth of Dr. Moreau," *Cahiers victoriens et edouardiens* 13 (1981), 31–41, 37; Mark M. Hennelly, "Reader Vivisection in *The Island of Dr. Moreau," Essays in Arts and Sciences* 9 (1980), 217–33, 217; and Lansbury, "Gynaecology," 435.
71. I agree with Bernard Bergonzi's reading that Moreau is both "a nightmarish caricature of the Almighty" and "a hypostatized image of the pretensions of science." Bernard Bergonzi, *The Early H.G. Wells* (Manchester: Manchester University Press, 1961), 106.
72. H.G. Wells, *The Island of Dr. Moreau: A Variorum Text,* ed. Robert M. Philmus (Athens, Georgia: University of Georgia Press, 1993), 10. Subsequent citations of this work will be indicated by parenthetical references in the text.
73. Roger Bowen has observed that, "in the fluid, metamorphic world of this novel, no one stays the same." Roger Bowen, "Science, Myth, and Fiction in H.G. Wells's *Island of Dr. Moreau," Studies in the Novel* 8 (1976), 318–35, 330.
74. Haynes, "Wells's Debt to Huxley," 40. See also Roslynn D. Haynes, "The Unholy Alliance of Science in *The Island of Dr. Moreau," The Wellesian* 11 (1988), 13–24, 16; and Elmar Schenkel, "Die verkehrte Insel: *The Tempest* und H.G. Wells's *The Island of Dr. Moreau," Anglia* 111 (1993), 39–58, 46.

75. Haynes, "Wells's Debt to Huxley," 33.
76. Haynes, "The Unholy Alliance," 13.
77. Haynes, "The Unholy Alliance," 19.
78. In the original draft of his novel, Wells gave Moreau a wife and assigned her a significant role in the work. His decision to edit her out of the published version suggests a rethinking of Moreau as a more self-oriented scientist. I thank Anne Stiles for calling my attention to the early draft of *Dr. Moreau*.
79. Wells also leaves open the possibility that Moreau is a sadist who is using science to channel his drives. I am grateful to Anne Stiles for pointing this out to me.
80. Sander L. Gilman, *Making the Body Beautiful: A Cultural History of Aesthetic Surgery* (Princeton: Princeton University Press, 1999), 3.
81. Haynes, "The Unholy Alliance," 22.

2

Our Lady of Darkness: Decadent Arts & the Magnetic Sleep of Magdeleine G.

Don LaCoss

> [Magdeleine G. was] *Mater Tenebrarum*, Our Lady of Darkness, the defier of God, the mother of lunacies and the suggestress of suicides, who "moves with incalculable motions, bounding, and with a tiger's leaps," who "wears the fierce light of a blazing misery that rests not for matins or for vespers, for noon of day or noon of night, for ebbing or for flowing tide." The dances of delight and gaiety come almost instinctively to those of a happy temperament; but the dances of grief, of fear, of madness, of despair, these are they which put the dancer to the severest test, which strip her of the acquired graces of the schools [of dance], and leave her dependent only on the quality of her own soul.[1]

Between 1904 and 1907, a dancer known professionally as "Magdeleine G." caused a stir in opera houses, theaters, private homes, medical clinics, and artists' studios across Central and Western Europe. Bourgeois spectators, journalists, eminent medical professionals, and *avant-garde* visual artists and musicians eagerly turned out to see the recitals, and some requested (and paid for) private performances: the Austrian ambassador hosted one such evening at his home in Munich, while in Paris, sculptor Auguste Rodin hired her to dance in his studios, supposedly saying afterward that her gestures were so "extremely beautiful" that "it is beyond what the spirit can conceive."[2] On stage, Magdeleine G.'s accompanists were frequently among the most prominent musicians of the day, such as German pianist and composer Bernhard Stavenhagen, who later went on to teach piano master classes at the Geneva Conservatory and conduct premiers by Gustav Mahler, Arnold Schönberg, and Claude Debussy. Gustave Doret, the conductor who had debuted Debussy's Impressionist tone poem *Prelude to the Afternoon of the Faun* (1894), played music for her London opening and wrote admiringly of her "most extraordinary and faithful translations of rhythm and musical thought."[3] Poets composed verses to her style and creativity while photographers and painters captured her portrait.

Comparisons between her art and that of Isadora Duncan and the Symbolist sensation Loïe Fuller were common; critic Ernst Schnur was one among many who declared that Magdeleine G. was, in fact, even more innovative and expressive than Isadora Duncan.[4] Reporters who attended her shows characterized her dancing as "stupefying,"[5] "the very embodiment of human passion,"[6] and "nearly savage in its intensity."[7] A reporter who had watched her London show called her "the most extraordinary, most impressive and most marvelous thing that we have seen,"[8] while one lout from the British daily *The Morning Leader* claimed that "she expresses more in just one gesture than the majority of women can in a half-hour conversation."[9]

A century later, though, Magdeleine G.'s importance and influence in the field of modern dance have been wholly eclipsed by Duncan, Fuller, and her other contemporaries. Perhaps what had made Magdeleine G. so famous in the early 1900s was too trendy to stand the test of time: she danced while unconscious. To be more precise, Magdeleine G. was a "magnetic subject," someone who had been put into a trance state that made him or her receptive only to the voice, touch, and influence of the mesmerist and who would later awake with no recollection of what had transpired during the somnambulistic episode.[10] Billed as the *Traumtänzerin* ("dream dancer") and the *Schlaftänzerin* ("sleeping dancer") in the press, her gimmick was that she was a dancer who lacked any musical or terpsichorean education but who, upon being put into a trance by her personal mesmerist (and promoter, stage manager, and business partner), transmogrified into a performance artist gifted with dance and dramatic mime faculties of the most modern sort.

One explanation for the notoriety of Magdeleine G.'s staged presentations of somnambulist dance can be understood by locating her spectacles within a field of specific modernist literary and artistic practices, conventions, and aesthetics evident in *fin-de-siècle* Europe, namely that of the Decadent arts movement. Linked to the modernist currents of Naturalism, Symbolism, and Expressionism, Decadence had emerged in a number of major European cities, starting in Paris in the early 1880s as a rebellion against the dominant liberal-bourgeois styles and values, especially self-serving notions of positivism and progress. In the arts, Decadence arose as a critical response to Realism, a revolt that many of its adherents grounded in theories of the unconscious mind's role in the production of artistic inspiration and creative intellect.

While there is no direct historical evidence to indicate that the shows were explicitly designed to trigger specific aesthetic associations for the audience that would help convince them of Magdeleine G.'s uncanny talents, a closer look at some of the musical and dramatic pieces selected by the show organizers (presumably including Magdeleine G. herself) reveals a preponderance of Decadent and Symbolist influence. Magdeleine G.'s unconscious creative energies were articulated as movement, and much of the discourse used to represent her subconscious powers drew on a preexistent Decadent sensibility that questioned the materialist values dominant in the urban centers of Imperial Germany and the Austro-Hungarian Empire before the outbreak of World War One—as one professor of ethnology pointed out not too long ago, part of Magdeleine G.'s popularity seemed to stem from the ways in which her performance radiated "an aura" of "decadence and liberty" for those who saw her.[11] For those who participated or who were familiar with the Decadent and Symbolist denunciations of Realism, Magdeleine G.'s somnambulistic performance art pointed the way toward a new means and form of creative expression.

What follows is my reconstruction of how the Decadents' artistic interests in the subconscious mind as a sustained, anguished interiority and a systematic derangement of the senses were associated with elements of Magdeleine G.'s stage routine. By so doing, I have excavated new sets of connections between neurology, literature, and modernism in the first decade of the last century. Though there is no direct evidence of deliberate, conscientious applications of Decadent aesthetics to her explorations and elucidations of the phenomena associated with unconscious mental processes, the resonances that I have located here point to how her dance exhibitions were wired into a cultural matrix of the broader innovations happening in Munich, Vienna, and the other cities where she performed.

The case of Magdeleine G.

In the spring of 1902, Magdeleine Guipet, a young married mother of two children, sought hypnotic psychotherapeutic treatment from neo-mesmerist Professor Émile Magnin of the Paris École de magnetisme for persistent headaches that she suspected were of neurasthenic origin. Magnin considered the possibility that she was a hysteric—he noted that, at age seven, Guipet had been exposed unexpectedly to a corpse, a standard event of deep psychical shock and trauma that was seen in late-nineteenth-century medical literature as an almost inevitable indicator of hystero-epileptic convulsions later in life. The initial diagnosis of hysteria was bolstered for Magnin by Guipet's medical history of color-blindness and debilitating migraines, her poor attention span, her recurring complaints of numbness that afflicted the left side of her body, her inability to sleep alone in a room or under direct moonlight, her fear of wind and thunder, and her neurotic unwillingness to walk to the right of anyone walking beside her.[12] Furthermore, there was the question of her racial predisposition to hysteria and atavism. Although hailing most recently from France, Guipet had been born in the Russian Empire in the Georgian city of Tbilisi in 1876 to a Swiss father and a Franco-Georgian mother, meaning that her bloodline was "mixed" and supposedly tainted by the irrationality, laziness, voluptuousness, and weak-mindedness that were common stereotypes of the Oriental "race."[13]

But by the fourth treatment session, Magnin had ruled out hysteria. He was fascinated by how Guipet had acted in response to the sounds of a metronome, music, and commanding voices while magnetized, her somatic reactions changing according to tone, pitch, volume, and duration. Magnin recruited a friend to play music during the next mesmeric session, and Guipet's unconscious response to the Chopin waltz came as a remarkable interpretive dance with bold and dramatic gestures that exuded precision, poise, and aplomb. When the magnetopath snapped her out of her somnambulistic state, Guipet claimed no memory of what had just happened. Guipet demonstrated a high degree of dance and dramatic mime talent while entranced, an observation that Magnin had confirmed by the director of the Opéra-Comique's ballet company who had attended the next therapy session. The director told Magnin that, although it was obvious that Guipet had talent, it was also clear that she had no professional training—in other words, her responses to music while deep in magnetic sleep were not hysterical convulsions as much as they were artistic movements and gestures that had sprouted fully formed from her unconscious mind and not from any remnants of academic dance training.[14] Was there some special neurological or psychological condition specific to Guipet that made her into such an accomplished dancer while mesmerized? Or was she proof of some greater, universal condition in humans, an example of how there was a murky, submerged reservoir of artistic talent within all individuals that could be tapped under hypnotic conditions that would make anyone a creative genius?[15]

Word of Magnin's experiments with Guipet (who appeared in print under the discreet patient pseudonym of "Magdeleine G.") traveled quickly among a certain segment of the European medical establishment who were investigating mediumistic phenomena and trance states as part of the larger debate over hypnotherapy. In 1903, the psychotherapist, criminologist, and sexologist Baron Albert von Schrenck-Notzing invited Guipet and Magnin to travel to Munich to meet with him; at the time, Schrenck-Notzing was well-known for importing to Austria and Germany the controversies over hypnosis and hysteria that were raging in French medical circles. Schrenck-Notzing arranged a private séance for Magdeleine G. and invited a number of doctors, artists, writers, and journalists to attend. The performance was interrupted when excited audience members began accusing Magnin, Magdeleine G., and Schrenck-Notzing of perpetuating a hoax: Magdeleine G.'s dancing struck many as being too accomplished to be the art of an unconscious amateur. In response to the outcry, Schrenck-Notzing prepared a special session for Magdeleine G. with members of the doctors' professional association.

The demand to see Magdeleine G.'s dream-dancing grew to the point that Schrenck-Notzing and his colleagues in a Munich paranormal investigations group needed to book a three-night run at the Schauspielhaus theater in Munich (Figure 2.1). Five thousand people paid 20 marks each to see Magdeleine G. dance and act out dramatic readings; the shows were such a critical success (and, at ten times the regular ticket price of a Schauspielhaus spectacle, presumably a financial one as well) that a seven-month tour of the opera house and theater stages of Vienna, Stuttgart, Paris, and London was arranged for Magdeleine G. and Magnin.

On each night, the shows began with Magdeleine G. sitting comfortably in a Louis XIV-style stuffed armchair wearing a long, sleeveless, light turquoise blue Classical-style peplos and soft-soled ballet shoes. Magnin, assisted by his wife Elsa, would execute a series of passes and put Magdeleine G. into magnetic sleep. He would then cue the musicians (often piano, viola, cello, flute, or voice) to begin and she would rise up from the chair—eyes open and fixed but sightless—and whirl, leap, and pose in response to a selection of music and dramatic readings. Each evening's music and dramatic selections were an unannounced combination of well-known works as well as improvisational pieces that were meant to eliminate any possibility of secret rehearsals on the part of Magdeleine G. and Magnin. One of the physicians who saw her in Munich said that she "accompanies the music and shouted suggestions [. . .] in a rather precise way, even according to pitch, volume, sound color, intervals and rhythms. If the music is suddenly interrupted, catalepsy follows: her

Figure 2.1 Photographer unknown, Émile Magnin and Magdeleine G. on stage at the Munich Schauspielhaus, March 1904. From Émile Magnin's *Art et l'hypnose. Interpretation plastique d'oeuvres littéraires et musicales* (1907), p. 263.

last motion freezes."[16] Audiences were forbidden to applaud for fear that the noise would shock Magdeleine G. awake, but, for the most part, their reactions were clearly enthusiastic: "Madeleine has unleashed storms of delight, she has set thousands and thousands in a state of euphoria," reported the modernist dramaturge Georg Fuchs, author of *Theater of the Future* (1905) and later a co-founder of the Munich Artists' Theater. Never had he seen all of the "human mysteries, the agony and the desire" of "Eve, Judith, Salomé, Clytemnestra, Helen, Antigone, and Cassandra" played so deftly by a single person in a single evening's show.[17]

Not everyone was as impressed as those who had been sitting with Fuchs. Reviewers and arts critics complained that Magdeleine G. was a phony, but some of the most vocal detractors of her performances were physicians, psychologists, neurologists, and medico-legal experts.[18] One of the most zealous critics of Schrenck-Notzing's experiments with Magdeleine G. was Dr. Albert Moll. In general, Moll did not object to medical hypnotherapy to treat those suffering from physical, psychological, and sexual difficulties—Moll and Schrenck-Notzing had been allies in the legal fight to criminalize the non-medical use of hypnosis by spiritualists, psychics, faith healers, and sideshow performers.[19] What upset Moll was Schrenck-Notzing's uncritical involvement with Magdeleine G. and her Svengali, Magnin, a sponsorship that struck Moll as a complete reversal of his colleague's principled stand against the sensationalized occultism and razzle-dazzle of fairground stage shows. "We know that people can dance, likewise we know that people in hypnosis can dance. I therefore see no scientific problem in the fact that Mme. M[agdeleine] dances in hypnosis," Moll grumbled.[20] He likewise complained:

> The way in which [Schrenck-Notzing] intervened in the case of Mme. Magdeleine G., the sleep-dancer, was calculated to make the public think there was something occultistic about her performances, and was very reprehensible. I am referring to his unjustifiable assertion that, in the first experiment, the effect of the music on the somnambulistic lady was such, that she developed a power of dramatic expression "far beyond the possibilities of the actor's art." Still more reprehensible was the way in which he foisted this lady, whose performance contributed nothing new to science, upon an unsuspicious public. In my opinion, the way he stage-managed the lady's performances in the presence of large audiences was an insult to science, and such methods should be rigidly excluded from the laboratory psychologist.[21]

Looking at certain aspects of Magdeleine G.'s 1904 Munich debut, Moll's concerns seem understandable. Schrenck-Notzing's presentation of Magdeleine G.'s magnetic sleep talents certainly came across more as spectacular entertainment than as a scientific examination. For example, the Munich Schauspielhaus that was booked for the performances had been recently designed by Jugendstil interior designer Richard Riemerschmid of the Munich United Workstation modern arts and crafts group. This new theater had a reputation as an experimental venue that offered challengingly Naturalistic fare such as Max Halbe's *Youth* (1893) and Gerhart Hauptmann's *The Beaver Coat* (1893), as well as Oscar Wilde's Decadent play, *Salomé* (1892).[22] Such a trendy setting for Schrenck-Notzing's sponsored appearance of the mesmerized sleep-dancer probably did little to convince skeptics of the sound scientific rationale for Magdeleine G.'s abilities.

Adding to the controversy was the enthusiasm that so many modern artists had for Magdeleine G., such as Hugo von Habermann. Habermann, part of the dissident modernist Munich Sezessionist movement, produced (sometimes sinisterly gynophobic) canvases of *femmes fatales* (Ophelia, Judith, Salomé), vaguely menacing sirens, and ecstatic bacchantes; his Jugendstil portrait *Magdeleine Guipet* (ca. 1904) appears to be a backstage dressing room, awash in murky browns, slate grays, and aquarium greens, wearing a foundation garment and being buffeted by unseen forces. Her eyes are closed, her back slightly arched, and her head is thrown back in a broad, otherworldly smile, suggesting that she is being depicted in a magnetic sleep.[23] Some strange material (her peplos?) appears to have taken on a life of its own and surges around her in tangles and twists like a medium's ectoplasm.

Another Sezessionist painter who had done at least 20 portraits of Magdeleine G. was Schrenck-Notzing's longtime colleague Albert von Keller. Blasted as a "Decadent" by German Roman Catholic authorities,[24] Keller had used entranced psychics (like the famous Eusapia Palladino and Lina Matzinger ["Lina M."]) and hypnotized women as models since the 1880s, believing them to be the most ideal subjects for modern artists whose art sought to visually represent the invisible realities of the mind and spirit.[25] "I have never encountered a soul so capable of domineering its body and able to give it such admirable expression in terms of intensity, beauty, and purity," Keller testified. "Magdeleine G. has the power of unlimited and inexhaustible expression for the beautiful."[26] Interestingly, he often painted the trance states of Magdeleine G. in the role of the mythological Trojan prophet Cassandra, a soothsayer whose predictions were absolutely accurate but who had been cursed by the god Apollo never to be believed and was condemned to be forever dismissed by those who heard her to be a madwoman.[27] Keller's portrait of Magdeleine G.'s Cassandra captures her in the throes of a magnetic fit, clad in her Classical dress with arms thrown open toward some unseen source of illumination into which she stares wide-eyed, face frozen in an ecstatic rictus of unknown origin, her left arm mysteriously dissolving into a nebula of misty light and shadow (Figure 2.2). Keller was also closely associated with the visual effects used in her Munich Schauspielhaus shows, as Schrenck-Notzing had hired him as the art director for the dream-dancer's 1904 performances.

Like Keller, Jugendstil portrait artist Friedrich August von Kaulbach also produced images of Magdeleine G. that Magnin reproduced in *Art and Hypnosis: Plastic Interpretation of Literary and Musical Works* (1907) to illustrate the findings of his experiments. Kaulbach, who painted a number of dancers, opera singers, and actresses, as well as aristocrats like Empress Elisabeth of Austria, viewed Magdeleine G. as a "marvelous apparition of beauty and inspiration."[28] Along with Keller and the famous Bavarian Symbolist painter Franz von Stuck, Kaulbach co-signed a statement of support for the *Traumtänzerin* that appeared in a Munich newspaper in 1904 as a response to some of the highly critical complaints published in the paper by medical observers like the aforementioned detractor Albert Moll.[29]

Figure 2.2 Albert von Keller, *Magdeleine Guipet as Cassandra.* From Oskar A. Müller, *Albert von Keller. 1844 Gais/Schweiz-1920 München* (München: Verlag Karl Thiemig), p. 84, pl. 104.

As we have seen, Magdeleine G.'s dancing straddled the dividing lines between science and art, between skepticism and the occult, and between the material world and the unconscious mind—as an example, consider the remarks of one reviewer who insisted that Magdeleine G.'s performance was a "curious artistic-scientific [*artistico-scientifique*] spectacle" while, the next day, another called it "a very curious scientific-artistic [*scientifico-artistique*] experience."[30] It was precisely this fuzzy shading of the differences between natural, preternatural, and supernatural phenomena that was the source of the controversy that surrounded her and excited both her fans and her disbelievers. The ambiguous and liminal nature of her art and its relationship to human nature is what made her such an unmistakable magnet for Decadent sensibilities.

Against nature

"Decadence" was the label adopted by a generation of writers (and later, painters, sculptors, and musicians) who were experimenting with a pessimistic style of philosophical anti-materialism that mixed internal drives with external forces. By the mid-1880s in Paris, manifestos and small periodicals like *Le Décadent* (1886–8) were being published to promote their ideas and work, prompting other literary and artistic cliques to activity in Brussels, St. Petersburg, London, and elsewhere throughout the Continent. This included Central and Northern Europe—from Vienna's Café Griensteidl to the Zum Schwarzen Ferkel wine bar on Berlin's Unter den Linden to the Grand Café in Kristiania (Oslo), small circles gathered to discuss the means to crack the hold of "the holy bourgeois trinity of morality, seriousness, and rationality."[31] Like the Symbolists they paralleled, cross-pollinated, and with whom were often confused, the Decadents honed a kind of anarcho-aristocratic solipsism that repudiated the arrogant, ignorant pretensions of the "cluster of symptoms called 'civilization.'"[32] Decadent creative productions asserted that the dominant bourgeois strategies of artistic and cultural representation were miserable half-measures that cheapened reality, so they responded with highly subjective atmospherics of exhaustion, hallucination, depravity, and melancholy meant to re-enchant life.

Aspects of this aesthetic can be found in Magdeleine G.'s stage shows in Vienna, Stuttgart, and Munich and in the language of the positive reviews of them. While there may very well be nothing more "natural" for individuals today than a plunge into the wilderness of intuition, instinct, and inspiration, Magdeleine G.'s magnetic sleep dance depictions of raw emotion, passion, and mysticism were understood a century ago as a radically stylized and unreal form of modernist idealism grounded in the experiences of the unconscious mind. Her somnambulistic dramatic and terpsichorean performances were the antithesis of Realism, instrumental reason, and tradition, making her art especially interesting to those intellectuals, writers, and artists who were sympathetic to the Decadent pursuit of the poison that would corrode bourgeois categories of the real and the natural.

Much of the Decadent perspective is an attack on the dominant role of materialism in framing middle-class realities in the late 1800s. Magdeleine G.'s mysterious talents contradicted the accepted contours of the material world; magnetopath Émile Magnin's explanations for her dancing fused the materialism of physical reality with the subjective truths of the unconscious, claiming that her dancing and gestures had to do with "hyperaesthetic" tactile sensory abnormalities that allowed her to feel the vibrations of music of which she was not aware in her waking state. In her magnetized state, Magnin speculated, Magdeleine G. could transform the sound vibrations of musical instruments into electric vibrations that her nervous system directly processed.[33] This enabled her to reflexively respond immediately and masterfully to improvised musical speeds and rhythms that one could not otherwise anticipate perceptually. The speed and rhythm of a piano piece, acting directly on her nervous system, triggered neurovascular effects that stimulated "an intensity of expression in all passions," biomechanical impulses that combined with "the creative power of her imagination, or with the many artistic, literary, and historical reminiscences that she draws from her subconscious" to find expression as dance.[34] To explore this theory, Magnin even conducted a number of outdoor performances in order to limit "structural interferences" on her nervous system that may have been occurring in clinics and acoustically engineered concert halls. "Where does psychology end? Where does physiology begin?" he asks rhetorically in his book.[35]

There is nothing to indicate that Magnin knew anything about those artists and writers who were experimenting with the neurologically influenced aesthetics of Decadence, Symbolism, Art Nouveau, and Jugendstil, but aspects of Magdeleine G.'s performances can be seen as dovetailing with a much larger search for a modern art of nerves in Central Europe.[36] Such experiments were motivated by a sense that the culture was in crisis, and all shared the urgent desire to dialectically surmount the material world and re-imagine nature, to commingle the dark subjective sensibilities of the poet with the scientist's harsh light of objective reason. The nervous system, or more suggestively, the unconscious, was the secret of this synthesis, a possibility that had been pursued by the Decadents and Symbolists in the decades before Magdeleine G. took the stage.

Goddesses of immortal hysteria[37]

Supposedly, Magdeleine G.'s unconscious powers were not determined by some specific genre of music or dramatic literature or its content; it was alleged that her subconscious mind was susceptible to all aural stimuli regardless of fashion. Therefore, it is entirely plausible that, in the interests of objective scientific inquiry, any kind of musical and dramatic material could have been used in her program: musicians could have limited their selections to well-worn concert standards, or played improvisational pieces to preclude any possibility of performance preparation on the part of the magnetopath and his subject, or perhaps random audience members could have been asked to pick the readings used for her dramatic interpretations. But instead, it seems that very calculated choices were made by the tour's organizers which specified the kind of material that would be used, and much of it had well-established attachments to the Decadent aesthetic.

For instance, there were all the Romantically idealized portrayals of tragically over-emotional, superhumanly passionate, or murderously insane women that figured in the majority of the musical and dramatic material chosen for Magdeleine G.'s repertoire. These historico-mythic *femmes fatales* included Judith of Bethulia and Delilah from the Hebrew Bible (the latter by way of Camille Saint-Saën's Romantic opera *Samson et Dalila* [1877]); Shakespeare's deranged Lady Macbeth; Gaetano Donizetti's mad and murderous *Lucia di Lammermoor* (1835) and *Lucrezia Borgia* (1840); and Oscar Wilde's *Salomé*. Some of these women had been reinterpreted by Decadents and Symbolists for a diverse number of cultural reasons at the *fin-de-siècle*, but none were more notoriously iconic than the latter, especially after Wilde's eponymously named and widely banned play, the final scene of which a mesmerized Magdeleine G. acted out on opening night at the Schauspielhaus.[38]

Magdeleine G. also mimed to recited verses of the *poète maudit* Paul Verlaine, whose work had been pressed into service by Symbolists and Decadents in the 1880s as a valuable precedent.[39] She almost always performed renditions of Frédéric Chopin's nocturnes and his "Funeral March," Romantic works that, like the poetry of Verlaine, had been reinterpreted to give it a tangible *fin-de-siècle* spin.[40] She danced to Romantic Franz Schubert's bleak, supernatural art-song *The Erlking* (1815), a retelling of Goethe's take on a nightmarish Nordic-Teutonic folk tale that surfaced at the *fin-de-siècle* in Russian Symbolism and in the music of Henri Duparc, a French Decadent composer who based his pieces on the morbid and melancholy poems by Baudelaire, Théophile Gautier, and the Symbolist Charles Leconte de Lisle. Quite a few of Richard Wagner's Romantic operas also figured into Magdeleine G.'s sleep-dances, including the pilgrim's chorus and bacchanal dance from *Tannhäuser* (1845), the prelude to *Lohengrin* (1848, a piece that Nietzsche called "the first example, only too insidious, only too successful, of hypnotism by means of music"), and the final ecstatic death scene of *Tristan and Isolde* (1859).[41] Five mystical selections culled from Wagner's musical stage-festival tetralogy, *The Ring of the Nibelung* (1854–74), were also included: "The Rhinemaidens' Song," "Ride of the Valkyrie," "Magic Fire Music," "Forest Murmurs," and "Siegfried's Funeral March."

The use of Wagner's music is worth a closer look in the context of the Decadent patterns in Magdeleine G.'s program. Though opera music had always trafficked in emotion, crimes of passion, and sensationalism, Wagner was accused by his opponents of raising the ante with tales involving overwrought passions blown up to a ritualistic scale and for portentously manufacturing entirely new forms of enervating mysticism that promoted unhealthy mythical archetypes, value systems, and existential structures of human experience.

Propagators of bourgeois moral panic feared that the content of Wagnerian opera had been deliberately tooled to appeal to the debased, sick, and degenerate segments of the population and thus cynically exploited their weaknesses and obsessions. Some critics of Wagner explained the poisonous nature of his compositions by pointing to the man himself as the cause of this pollution.[42] An ironical but obviously ambivalently troubled Nietzsche famously diagnosed him in 1888 as a modern nervous disease ("*Wagner est une névrose*") whose "sick," "convulsive," and "overexcited" music centers around the "problems of hysterics."[43] Social hygienist Max Nordau, meanwhile, charged Wagner with "a greater abundance of degeneration than all the degenerates put together" in the 600 pages of his study of the *fin-de-siècle* culture wars, referring to him as "the last mushroom on the dunghill of Romanticism."[44] His music-dramas provided "deep insight" into the "erotically emotional degenerate nature," Nordau howled. "Hysterical women were won over to Wagner chiefly by the lascivious eroticism of his music."[45]

The reputation of this music's mental, physical, or, at the very least, cultural toxicity partially accounts for its inclusion in Magdeleine G.'s show. When Magdeleine G. danced on stage to Wagner's music in Paris, Munich, and London in 1904, most in the audience would have been familiar with the rumors of manic nervous excitability and social decay that surrounded him. There were also associations being made at the time between Wagner's music and the latest theories of the nature and structure of the unconscious mind that emphasized formlessness, ineffable desires, and the seemingly unexpected return of repressed leitmotivs, especially in the poetico-speculative theorizations by the Symbolist and Decadent contributors found in publications like the *Revue wagnérienne* (1885–8) and *La Revue blanche* (1891–1903).

Magdeleine G.'s physical responses to Wagner's music elicited statements from critics that sound like they were lifted from a Mallarmé poem printed in the pages of the *Revue wagnérienne*. In the role of a grief-stricken Brünnhilde ordering the construction of Siegfried's pyre into which she will plunge to immolate herself and trigger a cosmic disaster that will destroy the gods, Magdeleine G. "rose to really superb heights. Such immensity in the sadness! One involuntarily thinks of how marvelous this woman who has such talent as mime could have been as Kundry [the tragic enchantress from *Parsifal*]," wrote the prolific turn-of-the-century music critic Hugues Imbert.[46] A critic in Munich agreed that she was "powerful and passionate" and that her "manner of interpreting Wagner was remarkable," but in the end he could not shake the unsettling impression that the event "was a serious and perhaps pathological affair."[47] One reviewer gushed that Magdeleine G.'s "glances, arms, bust, kneeling, immobility, quivering silences of the body, and ecstasies of the mute flesh" in response to the music of "Siegfried's Funeral March" could move Wagner himself from beyond the grave.[48]

From Wilde's Salomé to Wagner's Brünnhilde, Magdeleine G.'s complex attachments to playing the immortal goddesses of hysteria on stage were consistent with the anti-Realist aesthetic values promoted by Decadents. Given Magdeleine G.'s identification in the public mind and in medical circles as a practitioner of unconscious art, it is no coincidence that these stock characters showed up time and time again in her performances; Decadents and their Symbolist cousins had put forward the possibility that unconscious cerebration was the fountainhead of creative intellect in order to affect changes in the arts. The unleashing of the unconscious was expected to be a means of augmenting one's sensual array to create a new reality, a means of amplifying and magnifying the imagination as a force to create new, private worlds. Clearly, audiences were meant to infer for themselves the revolutionary modernist aesthetics that the unconscious mind can provide.

Drowning in the unconscious

"The Decadent analyzes felt emotions. He [*sic*] cultivates them with meditations upon his daydreams," an anonymous author spelled out in an article from the French periodical *Le Décadent* in 1888. In so doing, the Decadent "specifies them, seizing from them the most delicate nuances" through a systematic "folding up of thought, memory, and distinct images," substituting instead the most "vague and coarse primitive feeling" in order to explore the energies of the self.

> The universe is his laboratory . . . he chooses and rejects according to the exigencies of experimental groundwork or improvised testing. He is cruel without being coldly mechanistic, an irresponsible priest making the sacrifices commanded by the worship of his God.
> He wishes to know himself, he observes himself, he analyzes and notes everything down to the most evasive of half-felt emotions, the most tenuous quiver of psychic states scarcely yet formed, barely detectable; and by means of this deep, meticulous, remorselessly pursued investigation, he is constantly pushing back the frontiers of the unconscious.[49]

The motifs of dream, insanity, spiritual possession, religious and sexual ecstasy, communication with the dead, hypnosis, and intoxication (hashish, opium, morphine, ether, absinthe) that frequently appear in the Decadent arts are shorthand for these efforts to "push back the frontiers of the unconscious." But others were much more explicit about the ramifications of the latest theories of the unconscious for their aesthetic. The poet Jules Laforgue, for one, was so enthusiastic about philosopher Eduard von Hartmann's speculations on the unconscious that he wrote essays calling for the creation of a school of aesthetic responses to overthrow Realism. Historian, literary critic, Symbolist poet, and Decadent philosopher Rémy de Gourmont was likewise impressed and wrote the following passage in a review of Hartmann's book:

> We only know roughly according to which mechanism sensation is transformed into action. We only know that in order for that transformation to occur, the intervention of consciousness is unnecessary. We also know that, because of its power to modify determinist logic and to break a chain of free associations, such intervention can be harmful by inadvertently creating in the mind the first link in a new chain. Consciousness, then, which is the principle of liberty, is not the principle of art.[50]

This belief in the mind's unknown forces of innovative creative expression—a consciousness of the unconscious, in other words—percolated throughout commentary on Magdeleine G.'s dance recitals. One French-language arts critic found that quoting at length from the theories of Maeterlinck was the best way to explain the inexplicable in January 1904 after seeing the show.

> We have given names to its manifestations, we have called them instinct, soul, unconscious, subconsciousness, reflex action, presentiment, intuition, etc. We credit it more especially with the indeterminate and often prodigious force contained in those of our nerves that do not directly serve to produce our will and our reason: a force that would appear to be the very fluid of life. . . . Within us, underlying the conscious existence that our reason and will control, is a profounder existence, one side of which connects with a past beyond the record of history, the other with a future that thousands of years cannot exhaust. . . . Within us is a being that is our veritable ego, our first-born: immemorial, illimitable, universal, and probably immortal. Our intellect, which is merely a kind of phosphorescence that

> plays on this inner sea, has as yet faint knowledge of it. But our intellect is gradually learning that every secret of the human phenomena it has hitherto not understood must reside there, and there alone. This unconscious being lives on another plane than our intellect, in another world. It is familiar with all things; there is nothing it cannot do. . . . We do not detect any constant relation between the activity of the unconscious and the development of the intellect. This activity obeys rules of which we know nothing.[51]

Otto-Julius Bierbaum, co-founder of the lavish Symbolist and Decadent quarterly review *Pan* and author of the unwholesome, darkly sarcastic Decadent novel *Prince Cuckoo: The Life, Acts, Opinions, and Descent into Hell of a Libertine* (1908), initially had been dubious about Magdeleine G.'s show, but he soon was converted into a believer: "It had seemed to me that the thing had been too well-studied, because the artist's movements anticipated the sounds that they were to illustrate," he wrote in a Munich newspaper after viewing the first Schauspielhaus recital. But when he realized that she was dancing to music improvised on the spot, "I became convinced that this was a revelation of still unknown aesthetic forces." He continued to say that her performance was "of nearly uncanny force and of an absolute beauty. Never in my life have I seen such a way of bringing a human body" into expression. "It is only in the dreams that sometimes happen to us to dance like Magdeleine dances. It is only in dream, or when we are struck with rare inspiration." From the performance, Bierbaum concluded that "Beauty is not an invention of some privileged artists, but that it is something immanent within human nature itself."[52]

One of Bierbaum's collaborators at *Pan*, theater and literary critic Alfred Kerr, tapped a similar vein in his review of these biomagnetic trance-dances, saying that it recalled for him the mix of dark, compulsive desire, hypnotism, and hidden unconscious truths in one of the *Anatol* sketches (1892) by "the clever and ironical interpreter of Austrian Decadence," playwright Arthur Schnitzler.[53] Kerr also likened Madgeleine G.'s somnambulistic creativity to that of the infamous turn-of-the-century Decadent diva and Ibsenite, Eleanora Duse:[54]

> Is there an artist who creates while completely awake? I don't think so. Are we awake when writing? Do we not expunge 11/12ths of the world and dream up things that refer to one point only? Do you believe when watching a great, completely emotional actor that he does anything but sleep [hypnotic autosuggestion]? Do you believe

> that Duse is capable of achieving the loftiest effects, which grip the very seat of the soul, without sleep? Don't you believe it . . . Here, now, is an artist [Magdeleine G.] who has this ability of autosuggestion on a higher level . . . She is a strong performer of hysteria . . . but at the same time she must be a strongly hysterical performer . . . The inhibitions are swept aside (that's at least how it looks) and the innermost is turned outward.[55]

The affinity that Magdeleine G.'s exhibitions had for Decadent discourse was informed by the vast creative possibilities of unconscious mental activity. Adding to its appeal was the panicky resistance and outright disgust that many middle-class positivists had specifically directed at the existence of subconscious forces. Lamenting the efforts of the cutting-edge European literary modernists of his time, Max Nordau was one of those anti-Decadents who regarded the growing opinion that the unconscious was an unstoppable dynamo of creativity as one more symptom of Europe's degenerative cultural sicknesses, and those writers and artists who dared to capitalize on these degenerate tendencies were threats to the security of civilization.

> The "freedom" and "modernity," the "progress" and "truth" of these fellows is not ours . . . They wish for self-indulgence; we wish for work. They wish to drown consciousness in the unconscious; we wish to strengthen and enrich consciousness. They wish for evasive ideation and babble; we wish for attention, observations, and knowledge.[56]

The bourgeois terror of a deluge of unconsciousness that would swamp instrumental reason and drown civilization tied in with the catastrophist postures adopted by the Decadents. "I am the Empire as the decadence draws to a close," Verlaine famously penned in 1883, fashioning for decades afterward a formula for Decadents to use to liken themselves to Nero, fully satiated on violent melancholy splendors, indolently plucking at a lyre while society collapses all around them, passively watching the "Vandals' conquest."[57] In a sense, Magdeleine G. was a harbinger of such rack and ruin, a hysterical Cassandra whose visionary dreams were anxiously denied and repressed by those who, like Nordau, spurned the freedoms of the unconscious mind and instead endorsed a miserable regime of "work" and "consciousness" restricted by the most vulgar confines of empirical reality.

It was the understanding of the unconscious as a well-spring of unfathomed creative talent and expression that made Magdeleine G.

into a Decadent celebrity; the artistic abilities that roiled up out of her during magnetic sleep were recognized as evidence of the vital role of subconscious thought processes in the innovations of the creative intellect. Decadents, and those critics tangentially familiar with some of their theories and artistic practices, regarded the dream-dances of Magdeleine G. as an *avant-garde* illustration of how the unleashing of the unconscious could tune the senses and sensitivity to a higher pitch and force the imagination into new art and new worlds.

Acknowledgments

I must acknowledge the generous assistance provided to me by Katherine Pence, Gotje Anna Wäsc, and Patricia Zanski. Special thanks to my son Benjamin, a two-year-old Expressionist of great accomplishment.

Notes

1. J.E. Crawford Flitch, *Modern Dancing and Dancers* (Philadelphia: J.B. Lippincott, Co., 1912), 118. Flitch is quoting here from the "Levana and Our Ladies of Sorrow" section of Thomas De Quincey's *Suspiria De Profundis, being a Sequel to* The Confessions of an English Opium-Eater, which originally appeared in *Blackwood's Edinburgh Magazine,* 1844–5.
2. Quoted by Henri Carbonnelle in "Une séance d'hypnotisme chez Rodin," *Gil Blas* (27 November 1903), reprinted in Émile Magnin, *Art et l'hypnose. Interpretation plastique d'oeuvres littéraires et musicales* (Paris: Félix Alcan, 1907), 323–4.
3. Letter from Gustave Doret (Paris, 24 January 1905), reprinted in Magnin, *Art et l'hypnose,* 388.
4. Quoted by Fiona Trede in "Traumhaft Schön? Madeleine G. und der Tanz im Unbewussten" *Tanzdrama* 58.3 (2001), 6. This is also discussed in Albert von Schrenck-Notzing with Otto Schultze, *Die Traumtänzerin Magdeleine G.: Eine psychologische Studie über Hypnose und dramatische Kunst* (Stuttgart: F. Encke, 1904), 116–8. For more on how Magdeleine G.'s dancing was seen by reviewers of the day as more stirring and innovative than Isadora Duncan's, see Brygida Ochaim, "Franz von Stuck und der Tanz," in the exhibition catalog *Franz von Stuck und der Tanz. Der Tanz um die Jahrhundertwende* (Tettenweis: Förderkreis Franz von Stuck Geburtshaus Tettenweis e.V., 2001), 44–6.
5. Review from *Daily News* (4 May 1904), in Magnin, *Art et l'hypnose,* 375.
6. Review from *Daily Express* (4 May 1904), in Magnin, *Art et l'hypnose,* 375.
7. Review from *Daily Telegraph* (4 May 1904), in Magnin, *Art et l'hypnose,* 375.
8. Review from *The Star* (4 May 1904), in Magnin, *Art et l'hypnose,* 376.
9. Review from *The Morning Leader* (4 May 1904), in Magnin, *Art et l'hypnose,* 376.
10. The earliest discussion of the "magnetic subject" is by the Marquis de Puységur (Armand-Marie-Jacques de Chastenet) and his work on animal

magnetism in France in the 1780s. For more on the "magnetic subject" within the context of the debates around subjectivity and the unconscious, see Bertrand Méheust, *Somnambulisme et médiumnité, 1784–1930. Tome I: Le défi du magnétisme* and *Tome II: Le Choc des sciences psychiques* (Le Plessis: Éditions Synthélabo, 1991).

11. Clara Gallini, *La sonnambula meravigliosa. Magnetismo e ipnotismo nell'Ottocento italiano* (Milan: Feltrinelli, 1983), 107.
12. Magnin, *Art et l'hypnose*, 152–63.
13. Those who argued in favor of theories of racial degeneration in order to explain Guipet's talents pointed out that her maternal uncles were well-known Caucasian folk dancers and offered that her abilities could be attributed to the return of some repressed "racial memory." This diagnosis is dismissed in Émil Magnin, "Catalepsie et extase musicale," *Revue spirite* (July 1903), 3–10. For more on the primitivist and Orientalist racial stereotypes of Circassians in the nineteenth century, see Susan Layton, *Russian Literature and Empire: Conquest of the Caucasus from Pushkin to Tolstoy* (Cambridge: Cambridge University Press, 1994); Linda Frost, *Never One Nation: Freaks, Savages, and Whiteness in US Popular Culture, 1850–1877* (Minneapolis: University of Minnesota Press, 2005), 56–85; and Robert Bogdan, *Freak Show: Presenting Human Oddities for Amusement and Profit* (Chicago: University of Chicago Press, 1988), 235–41.
14. Many who challenged Magnin's findings did so by insisting that Guipet was a hysteric; in one instance, some pointed to the similarities between the gestures she made in reaction to the music of ecstatic transfiguration in Isolde's death scene (from Richard Wagner's *Tristan und Isolde*) to the classic "*arc de cercle*" cataloged by Jean-Marie Charcot in his studies of hysterics. For Magnin's response, see *Art et l'hypnose*, 152–63.
15. Three decades and one World War later, the surrealist poet André Breton acknowledged the value of Magdeleine G.'s dances in his influential 1933 essay calling upon readers to dig down into the unconscious mind in order to become mediums themselves; see Breton, "Le Message automatique," *Minotaure* 3–4 (December 1933), 41–64, 54. Breton discusses the reports made by Theodor Lipps and Alfred Schrenk-Notzing on the sources of Magdeleine G.'s talents; Marguerite Bonnet, in her notes to "Le Message automatique" in Andre Breton, *Œuvres complètes*, tome II (Paris: Gallimard, 1992), 1541, suspects that Schrenck-Notzing's *Die Traumtänzerin Magdeleine G.* might be the source for the quotes used by Breton. However, I have discovered that the comments by Lipps and Schrenck-Notzing used by Breton in "Le Message automatique" appear word for word (and in French) in Theodore Flournoy, "Chorégraphie somnambulique. Le cas de Magdeleine G.," *Archives de psychologie* 3 (July 1904), 357–74, 364n.1 and 373, respectively.
16. Dr. S. Seif, *Münchner Neueste Nachrichten* (1905), translated in full by Henry Marx in "Madeleine: Two Reviews," *The Drama Review* 22.2 (June 1978), 29–31, 30. The changes here to Marx's translation are my own.
17. Georg Fuchs, *Sturm und Drang in München um die Jahrhundertwende* (Munich: Verlag Georg D.W. Callwey, 1936), 241.
18. Magnin broke with Schrenck-Notzing during the European tour, supposedly because Schrenck-Notzing was not enough of a believer in magnetic sleep and insisted on retaining the anti-mesmerist categories of Braidian hypnosis.

19. Heather Wolffram, "On the Borders of Science: Psychical Research and Parapsychology in Germany, c. 1870–1939," (Ph.D. diss., University of Queensland, 2005), 81–125. I am grateful to Heather Wolffram for sharing with me her ideas and research on Schrenck-Notzing.
20. Albert Moll, *Psychologie und Charakterologie der Okkultisten* (Stuttgart: Ferdinand Enke, 1929), 20, qtd. in Wolffram, 124.
21. Albert Moll, *Hypnotism, including a study of the chief points of psycho-therapeutics and occultism,* trans. Arthur F. Hopkirk (London: Walter Scott, 1909), 533–4. For an outspokenly critical (and quite sarcastic) account of what happened at this performance, see the pamphlet by Dr. Franz Roberts, *Die Schlaftänzerin Madeleine G.; ein Protest gegen den Missbrauch der Wissenschaft* (Munich: n.p., 1904). Heather Wolffram has suggested to me the possibility that "Franz Roberts" was one of the many pseudonyms used by Moll in his denunciations of Schrenck-Notzing.
22. Peter Jelavich, *Munich and Theatrical Modernism: Politics, Playwriting, and Performance, 1890–1914* (Cambridge: Harvard University Press, 1985), 151–60.
23. A reproduction of Hugo von Habermann's *Madeleine Guipet* can be seen in the art journal *Cicerone, halbmonatsschrift für künstler, kunstfreunde und sammler* XI (1919), plate 12. My thanks to Sabine Scheele for her help with Habermann's work.
24. Corinna Treitel, *A Science for the Soul: Occultism and the Genesis of the German Modern* (Baltimore: Johns Hopkins University Press, 2004), 110–4.
25. Schrenck-Notzing argued that her principal value to the artist was as a type of "human still life." Schrenck-Notzing, "Albert von Keller als Malerpsychologe und Metapsychiker," *Psychische Studien* 48 (1921), 201–10, 208.
26. Letter from Albert von Keller (Munich, 18 December 1904), in Magnin, *Art et l'hypnose*, 393.
27. It is not at all clear if Magdeleine G. ever actually performed a piece involving Cassandra, such as acting out a scene from Aeschylus or dancing to the music of Hector Berlioz's *Les Troyens* (1858). Perhaps Keller was simply fascinated by the myth and the image of a madwoman who was ignored but who actually knew the truth. Georg Fuchs compares Magdeleine G. to Cassandra in *Sturm und Drang*, 241.
28. Letter from Friedrich August von Kaulbach (Munich, n.d.), in Magnin, *Art et l'hypnose*, 392.
29. The statement appeared in *Allgemeine Zeitung* (23 March 1904), in Magnin, *Art et l'hypnose*, 375.
30. "Artistic-scientific" comes from an article in *Corresponance Havas* (26 January 1904); "scientific-artistic" comes from one in *Journal de Rouen* (27 January 1904), in Magnin, *Art et l'hypnose*, 342–3.
31. Malcolm Green, introduction to *The Golden Bomb: Phantastic German Expressionist Stories* (Edinburgh: Polygon, 1993), 1–27, 9. For more on German Decadence, see Ray Furness's untitled introduction to *The Dedalus Book of German Decadence: Voices of the Abyss* (New York: Hippocrene, 1994), 9–25; Florian Krobb, "'*Die Kunst der Vater todtet das Leben der Enkel*': Decadence and Crisis in *Fin-de-Siècle* German and Austrian Discourse," *New Literary History* 35.4 (2004), 547–62; and Robert Vilain, "Temporary Aesthetes: Decadence and Symbolism in Germany and Austria," in Patrick McGuinness, ed., *Symbolism, Decadence and the* Fin-de-Siècle (Exeter: University of Exeter Press, 2000), 209–24.

32. Walter Benjamin, "On Some Motifs in Baudelaire," in *Charles Baudelaire: A Lyric Poet in the Era of High Capitalism* (New York: Verso, 1997), 107–54, 131.
33. Magnin, *Art et l'hypnose*, 179–80.
34. Magnin, *Art et l'hypnose*, 192–5.
35. Magnin, *Art et l'hypnose*, 215–6.
36. Similar arguments about the fusion of medicine and art are made in Debora L. Silverman, *Art Nouveau in Fin-de-Siècle France: Politics, Psychology, and Style* (Berkeley: University of California Press, 1989).
37. "[Painter Gustave Moreau's Salomé] is the symbolic incarnation of undying Lust, the Goddess of immortal Hysteria, the accursed Beauty, exalted above all other beauties by the catalepsy that hardens her flesh and steels her muscles; the monstrous Beast, indifferent, irresponsible, insensible, poisoning, like the Helen of ancient myth, everything that approaches her, everything that sees her, everything that she touches." Joris-Karl Huysmanns, *Against Nature*, trans. Robert Baldick (New York: Penguin, 1959), 66.
38. Dr. Fr[iedrich] Maier, "Die 'Schlaftänzerin' Madeleine G. in München," *Psychische Studien* 31.5 (1904), 239–46, 245. The scene from Wilde's play was read by Lili Marberg, an actress recognized early in her career for her modernist performances of Ibsen and Shaw.
39. Philip Stephan, *Paul Verlaine and the Decadence, 1882–90* (Manchester: Manchester University Press, 1974).
40. The Decadence of Chopin's music is alluded to in Wilde's *The Picture of Dorian Gray* (1890; revised 1891) and in Thomas Mann's *The Buddenbrooks: The Decline of a Family* (1901) and *Tonio Kröger* (1903); see William Eickhorst, *Decadence in German Fiction* (Denver: Alan Swallow, 1953), 33–4.
41. Friedrich Nietzsche, from Section 7 of *The Case of Wagner. A Musician's Problem*, in *The Birth of Tragedy and The Case of Wagner*, trans. Walter Kauffmann (New York: Vintage, 1967), 171.
42. Wagner's vile racism does not seem to figure in most of these analyzes directly; a more contemporary discussion of the subject can be found in Marc A. Weiner, *Richard Wagner and the Anti-Semitic Imagination* (Lincoln: University of Nebraska Press, 1995).
43. Nietzsche, from Section 5 of *The Case of Wagner*, 166.
44. Max Nordau, *Degeneration* (New York: D. Appleton and Company, 1902), 171. As a serious study of turn-of-the-century European arts, *Degeneration* is largely useless, but intellectually his views are completely consistent with prevailing liberal opinion of his day: a smug faith in scientific materialism, a commitment to the illusions of progress, a devotion to the moral importance of self-disciplinary repression, and a horror of contagion and cultural entropy. See George Mosse, "Max Nordau and His 'Degeneration.' An Introduction to Max Nordau," in *Degeneration* (Lincoln: University of Nebraska Press, 1993), viii–xxxiii.
45. Nordau, *Degeneration*, 190, 211.
46. H[ugues] Imbert, "De l'interprétation musicale dans l'hypnose," *Le Guide Musical* (31 January 1904), in Magnin, *Art et l'hypnose*, 357–8.
47. Alfred von Mensi, *Allgemeine Zeitung* (20 Feb 1904), in Magnin, *Art et l'hypnose*, 360.
48. O'Divy, "La musique et l'hypnose," *Le Soleil* (26 January 1904), in Magnin, *Art et l'hypnose*, 334.

49. "R.V.," "Un article sur les Décadents," *Le Décadent. Revue littéraire bi-mensuelle* 3.25 (15–31 December 1888), 14.
50. Rémy de Gourmont, "La Création subconscience," reprinted in *La Culture des idées* (Paris: Mercure de France, 1900), 41–5, 44. Philosopher Eduard von Hartmann's *Philosophie des Unbewussten* (1869) was translated as *Philosophie de l'inconscient* (Paris: Ballière, 1877) by D. Nolen. For more on Laforgue's and Gourmont's readings of Hartmann, see Jean Pierrot, *The Decadent Imagination, 1880–1900*, trans. Derek Colman (Chicago: University of Chicago Press, 1981), 121–2. In addition to Hartmann, the Symbolists and Decadents interested in creativity and the (pre-Freudian) unconscious were also taken with Dr. Paul Chabaneix's *La Physiologie cérébrale. De l'Inconscient chez les artistes, les savants, et les écrivains* (1897), and philosopher-psychiatrist Théodule Ribot's *Essai sur l'imagination créatrice* (1900); see Jeremy Stubbs, "Between Medicine and Hermeticism: 'The' Unconscious in *Fin-de-Siècle* France," in *Symbolism, Decadence and the* Fin-de-Siècle, ed. Patrick McGuinness (Exeter: University of Exeter Press, 2000), 167–72.
51. Maeterlinck, qtd. in O'Divy, "La musique et l'hypnose," *Le Soleil* (26 January 1904), in Magnin, *Art et l'hypnose*, 335. The text is from Maeterlinck's "La Chance" from *Revue de Paris* (15 March 1901), 241–66; translation here is by Alfred Sutro from Maeterlinck, *The Buried Temple* (London: Ruskin House, 1902), 250–1.
52. Otto-Julius Bierbaum, from an unidentified article in *Münchner Neuesten Nachrichten* (19 February 1904), qtd. in Magnin, *Art et l'hypnose*, 334.
53. Schnitzler is described in this way in Eickhorst, *Decadence in German Fiction*, 75.
54. By most accounts, Duse was "discovered" as an actress by Hermann Bahr during his Decadent tour to St. Petersburg in 1891; see Bahr, "Die Duse," in *Studien zur Kritik der Moderne* (Frankfurt: Rütten & Loening, 1894), 251–7. The following is from a review of Duse in an 1896 production of Dumas's *The Lady of the Camellias* in a New York newspaper: "Duse is not a natural actress nor an actress of technique . . . she is simply a nervous, magnetic woman who affects other nervous women and some nervous men. She is to stage what Madame Blavatsky was to religion. She is lauded to the skies for her realism, but she is truly the least realistic of actresses." Qtd. in Eva Le Gallienne, *The Mystic in the Theatre: Eleonora Duse* (Carbondale: Southern Illinois University Press, 1973), 133. For more on Duse in this vein, see the memoirs of Symbolist Aurélien Lugné-Poë, *La Parade sous les étoiles. Souvenirs de theâtre, 1902–1912* (Paris: Gallimard, 1933) and Decadent Arthur Symons, *Eleonora Duse* (London: Ayer, 1927). Schrenck-Notzing makes a Duse comparison as well; see Schrenck-Notzing, *Die Traumtänzerin Magdeleine G.*, 71.
55. Kerr (8 February 1905), in Marx, "Madeleine: Two Reviews," 29–31, 31.
56. Nordau, *Degeneration*, 556–60.
57. From the poem "Langueur" that originally appeared in the collection *Jadis et naguère* (1885), translated as "Languor" by Norman R. Shapiro in *One Hundred and One Poems by Paul Verlaine: A Bilingual Edition* (Chicago: University of Chicago Press, 1999), 135.

II. Diagnostic Categories

3
How Do I Look? Dysmorphophobia and Obsession at the *Fin de Siècle*

Andrew Mangham

> Millions of Americans have a secret obsession. They're obsessed with how they look, with a perceived flaw in their appearance. They worry that their nose is too big, their breasts are too small, their skin is blemished, their hair is thinning, their body build is too small – any body part can be the focus of this obsession. . . . Most of us care about how we look – we think about our appearance and try to improve it. . . . While the concerns of body dysmorphic disorder (BDD) echo these normal concerns, they're more extreme. People who have BDD not only dislike some aspect of how they look, they're *preoccupied* with it.
>
> Katharine A. Phillips, *The Broken Mirror: Understanding and Treating Body Dysmorphic Disorder* (1986)[1]

In 1986 psychiatrist Katharine Phillips introduced this definition of Body Dysmorphic Disorder (BDD). The condition, she explains, is a type of Obsessive–Compulsive Disorder where one experiences crushing preoccupations with one's perceived inability to live up to certain standards of beauty. Not to be confused with vanity, BDD, according to Phillips, is an obsession with a physical flaw that may or may not exist and which the sufferer sees as making him or her uglier than the average person.

The latter half of the twentieth century was a poignant time for the idea of BDD to emerge. Cosmetic surgery was, and still is, on the increase and this is no doubt due to the fact that, since the 1980s, both men and women have been under enormous strains to look as good as people in the media. It is in response to this pressure, suggests Phillips, that BDD has been identified as a psychiatric condition. As media pressures to look attractive have multiplied, so too have the number of people suffering from a debilitating sense of their own ugliness. Psychiatry has responded to these developments with new ways of identifying and treating BDD.

It is not the aim of this chapter to question the links between media pressures and the prevalence of BDD. What I aim to do is show how the condition actually has its origins in the late-nineteenth century. It was at this time, for example, that theories of "dysmorphophobia" and obsession were first articulated and these classifications responded, like BDD, to enormous cultural anxieties to look good. It is possible to trace the dysmorphic idea through a range of writings, both literary and scientific, in the late-Victorian period. Although originating in the writings of psychologists like Jean Esquirol in the 1830s, the idea of being obsessed with one's appearance was appropriated and modified by fictional narratives throughout the nineteenth century. In what follows I argue that the coinage of "dysmorphophobia" in 1891 owes a great deal to earlier fictional narratives and neurological models of consciousness propounded by mid-Victorian psychologists. Looking at some of these in detail allows one to understand how such responses to pressures to look good were part of the complex and protean discussions of descent and degeneration at the *fin de siècle*.

Following the 1859 publication of Darwin's *Origin of Species*, the nineteenth century developed a growing obsession with human breeding and healthy progeniture. The pressures of marrying well had been extant for centuries, but it was the Victorians who converted these issues into a scientific vocabulary. In 1865, for instance, Francis Galton, Darwin's cousin, contributed to *Macmillan's Magazine* with a couple of essays on "Hereditary Talent and Character." The essays anticipated the publication of his *Hereditary Genius* (1869) and contained a suggestion that, in order to maintain a healthy familial stock, one needs to select one's spouse carefully. He delivered a utopian vision of what the world would be like if men, in particular, were to make their conjugal choices with more care:

> What an extraordinary effect might be produced on our race, if its object was to unite in marriage those who possessed the finest and most suitable natures, mental, moral, and physical! . . . An examination has ... been conducted on established principles among all the young ladies of this country who are now of the age of twenty-one, and I need hardly remind you, that this examination takes note of grace, beauty, health, good temper, accomplished housewifery, and

> disengaged affections, in addition to noble qualities of heart and brain. . . . If a twentieth part of the cost and pains were spent in measures for the improvement of the human race that is spent on the improvement of the breed of horses and cattle, what a galaxy of genius might we not create![2]

Galton did not appear to be responding to any specific anxieties in this passage; at the *fin de siècle*, for instance, such arguments were designed to stop the perceived degeneration of the human race. Here Galton optimistically envisions a world improved through a more careful consideration of which individuals are best suited to marry and procreate. He chooses to focus entirely on women as though they are stock in a cattle market and does not concentrate solely on their looks: "Noble qualities of heart and brain," he suggests, are as important as grace, beauty, and health. Despite concurrent fears that women's appearances could not be trusted, Galton was writing at a time when noble and moral qualities were thought to be discernible through physical appearance. In the 1860s, for example, E.S. Dallas submitted a couple of articles to *The Cornhill Magazine* with the intention of creating a resurgence of interest in physiognomy – the "science" that claimed one could read a man's character by studying his face.[3] Dallas claimed that the invention of photography was responsible for a renewed interest in physiognomy yet, as numerous studies have shown, physiognomic principles were employed in fiction throughout the nineteenth century.[4] Although scientists did not take physiognomy seriously in the latter half of the nineteenth century, the notion of judging a character by his or her appearance grew in influence, taking on new signification in relation to pathology. Andrew Smith, for instance, discusses how the medical establishment claimed, at the end of the century, that signs of immorality and consequent ill sexual health manifested themselves in physical appearance. Syphilis, he notes, was thought to cause bodily scars – "signs" that could be read by doctors and ordinary members of the public.[5] This act of reading the body, adds Sander L. Gilman, developed into "syphilophobia," a cultural obsession with spotting the so-called indicators of sin.[6]

In the Victorian times, then, one's appearance needed to be "satisfactory" if one was to be considered as worthy for marriage. Outward bodily scars were considered to signify whether or not one was a healthy breeder. Such ideas were not exclusively linked to venereal disease either. *Any* tokens of illness were, it seems, looked upon as weaknesses that inevitably affected one's suitability for marriage. Concerns over the health of future generations, combined with the perceived possibilities of biological conditioning, led to the emergence of certain sexual ideals and the concept of possessing good health was a major factor among these. "The social and medical emphasis on woman as reproductive vehicle," writes Sally Shuttleworth, "heavily influenced social perceptions of female beauty and marriageability."[7] And, indeed, such pressures were not limited to women. As Jane Wood observes, the Victorians learned to appreciate the "virtues of a more vigorous and outgoing ideal of manliness."[8] Any deviations from these respective feminine and manly ideals were read as flaws and signs that the individual in question would make a poor contribution to any gene pool.

In Charlotte Brontë's *Jane Eyre* (1847) and Charles Dickens's *Bleak House* (1852–3), Victorian preoccupation with looks is portrayed as influencing an individual's sense of his or her own value. Jane Eyre admits:

> It was not my habit to be disregardful of my appearance, or careless of the impression I made: on the contrary, I ever wished to look as well as I could, and to please as much as my want of beauty would permit. I sometimes regretted that I was not handsomer: I sometimes wished to have rosy cheeks, a straight nose, and small cherry mouth; I desired to be tall, stately, and finely developed in figure; I felt it a misfortune that I was so little, so pale, and had features so irregular and so marked. And why had I these aspirations and regrets? It would be difficult to say: I could not then distinctly say it to myself; yet I had a reason, and a logical, natural reason too.[9]

The last comments in this extract are interesting in light of the growing pressures faced by Victorian women to look good. Jane's "logical [and] natural" reason for worrying about her appearance is that she suspects she needs to have rosy cheeks, a straight nose, and a finely developed figure in order to be considered attractive to the opposite sex. Yet pinpointing the exact cause of her self-torture is, as she claims, "difficult to say." This is because, while body-dysmorphic anxieties existed prior to the late-nineteenth century, the vocabulary for identifying them did not. Jane can describe her feelings of physical inadequacy, but it would take another 40 years before psychiatrists labelled her condition as "dysmorphophobia."

Later in the narrative, Jane's unhealthy feelings about her appearance become central to her mental torment. She compares herself to Blanche Ingram even before she has met her. She says to herself:

> Listen, then, Jane Eyre, to your sentence: to-morrow, place the glass before you, and draw in chalk your own picture, faithfully; without

> softening one defect: omit no harsh line, smooth away no displeasing irregularity; write under it, "Portrait of a Governess, disconnected, poor, and plain."
>
> Afterwards, take a piece of smooth ivory . . . delineate carefully the loveliest face you can imagine; paint it in your softest shades and sweetest hues, according to the description given by Mrs. Fairfax of Blanche Ingram . . . .
>
> Whenever, in future, you should chance to fancy Mr. Rochester thinks well of you, take out these two pictures and compare them: say, "Mr. Rochester might probably win that noble lady's love, if he chose to strive for it; is it likely he would waste a serious thought on this indigent and insignificant plebeian?"
>
> (161)

It is interesting how Jane's imagination has full reign in this episode. The fact that she has never met Blanche, and is yet convinced of her all-conquering beauty, is a good indication that Jane's anxieties are the product of her own unhealthy feelings. Indeed, later events reveal that Rochester is actually more interested in Jane.

Esther Summerson, the narrator of Charles Dickens's *Bleak House*, experiences a similar psychological crisis after suffering from a bout of smallpox. The scars left on her face, she believes, disqualify her from marrying the young doctor Allan Woodcourt. She learns from Woodcourt's mother that forming a good alliance is crucial to her son's pedigree: "Mrs. Woodcourt, after expiating to us the fame of her great kinsm[e]n, said that no doubt, wherever her son Allan went, he would remember his pedigree, and would on no account form an alliance below it."[10] Despite the fact that Woodcourt shows no complicity with his mother's feelings, Esther assumes that her scars will be read by him as marks of weakness and he will see her as unsuitable for wedlock. She believes that it is more charitable to release him from any attachment they may have formed. She narrates: "What should I have suffered, if I had had to write to him, and tell him that the poor face he had known as mine was quite gone from me, and that I freely released him from his bondage to one whom he had never seen!" (455). She then outlines the moment she first saw her mirror image after the illness:

> For I had not yet looked in the glass, and had never asked to have my own restored to me. I knew this to be a weakness which must be overcome . . . . [I] went up to the glass upon the dressing-table. There was a little muslin curtain drawn across it. I drew it back: and stood for a moment looking through such a veil as my own hair, that

> I could see nothing else. Then I put my hair aside, and looked at the reflection in the mirror; encouraged by seeing how placidly it looked at me. I was very much changed – O very, very much. At first, my face was so strange to me, that I think I should have put my hands before it and started back, but for the encouragement I have mentioned.
>
> (457)

What is important about Esther's changed appearance is the way the narrative foregrounds her own feelings about it. Notice, for example, her reluctance to look into the mirror. This reaction to one's mirror image is, according to Phillips, a textbook symptom of BDD. People with the condition have unhealthy relationships with mirrors; convinced of their own ugliness, they avoid looking into them.[11] When they do catch sight of themselves, they see a neurotic exaggeration rather than a reality. What are viewed as small or insignificant flaws by others are perceived, by the BDD sufferer, to be devastating deformities. Esther Summerson's scars are of course real. Her friends and family hide the mirrors to prevent her being upset by the change in her appearance. Yet, while her scars are not the fabrication of a neurotic obsession, her reaction to them is exaggerated. Without consulting Woodcourt, she assumes that he will no longer love her because she is scarred. When she does finally meet him, she discovers how, actually, her "scarred face was all unchanged to him" (754).

Such is also the case for Jane Eyre. Her impression that her plainness makes her unsuitable for Rochester's hand is not supported by the man himself, who values her for more than just her appearance. He says: "I would not – I could not – marry Miss Ingram. You – you strange – you almost unearthly thing! – I love as my own flesh. You – poor and obscure, and small and plain as you are – I entreat to accept me as a husband" (254–5). Jane's self-esteem, like Esther's, is improved greatly by the knowledge that someone finds her attractive despite her imperfections. The next morning Jane looks into her mirror:

> While arranging my hair, I looked at my face in the glass, and felt it was no longer plain: there was hope in its aspect, and life in its color; and my eyes seemed as if they had beheld the fount of fruition, and borrowed beams from the lustrous ripple. I had often been unwilling to look at my master, because I feared he could not be pleased with my look; but I was sure I might lift my face to his now, and not cool his affection by its expression.
>
> (257)

Notice the change in Jane's reaction to her image. Whereas before she saw herself as "plain" and "insignificant," she now sees hope, color, and luster. She has gone through no physical transformation, yet her thoughts about her appearance *have* changed. This suggests that they were never fixed or based on any reality to begin with. Instead, they were illusionary and exaggerated because driven by a neurotic reaction to cultural pressures. Such is supported by the fact that Rochester, like Woodcourt with Esther, does not concur with her assumption that she is too plain and insignificant for him.

This lack of correspondence between reality and the perceptions of the individual is typical of the dysmorphic obsessive. Whether they are real or not, Esther's disfiguring marks and Jane's plainness are internalised and brooded over. Brontë and Dickens present their readers with a view of how contemporaneous notions of marital worth were converted, by some insecure individuals, into unhealthy self-impressions.

The term BDD was not, of course, available to Brontë and Dickens when writing *Jane Eyre* and *Bleak House*. Yet, from the 1830s onwards, psychiatrists were developing neurological models based on the notion of fixed ideas or, to use the more modern term, obsession. In 1838 Jean Esquirol drew on the writings of Philippe Pinel to develop the idea of "monomania." According to Esquirol, this was a form of insanity directed towards a single, unhealthy object or impression. In his influential *Mental Maladies* (1838), he wrote: "The attention exercises itself with so much energy as to become fixed upon a single subject. Constantly confined to this, nothing can turn it aside. All reasonings and determinations are derived from this all-absorbing idea."[12] Clearly, Jane Eyre's and Esther Summerson's feelings are not as extreme as those outlined here. *Jane Eyre* and *Bleak House* present two characters who have unhealthy impressions of their own looks, but portray neither as driven to obsessive or extreme forms of behavior (despite the links modern criticism has drawn between Jane and Bertha Rochester). Later in the century, however, psychologists drew on the writings of Esquirol and his predecessors to develop ideas of thought as something that ran in fixed channels. Thinkers like G.H. Lewes, William Carpenter, and Herbert Spencer contributed to the field of physiological psychology. They believed that an individual's experiences and impressions were carved into the physical structure of the brain. This ongoing reshaping of the organ would, they suggested, influence future behavior. In *Problems of Life and Mind* (1874–9), for instance, Lewes wrote:

> The evolution of Mind is the establishment of definite paths: this is the mental organisation, fitting it for the reception of definite impressions,

> and their co-ordination with past feelings. . . . Through their registered modifications, feelings once produced are capable of reproduction; and must always be reproduced, more or less completely, whenever the new excitation is discharged along the old channels.[13]

Drawing on the controversial, earlier claims of both William Lawrence and Charles Darwin, such ideas are a result of Victorian scientific developments that witnessed neurology exert a growing influence over the discipline of psychology.[14] It was understood that psychological phenomena left a physical impression on the brain resulting in a complex, somatic evolution of the mind.

The neurological channels identified by Lewes were considered to be a strength in healthy minds. When behavior became habitual, it had carved, through repeated application, a channel in the brain, and future actions simply traversed the same pathway. Hence, one did not need to keep relearning the same skills, and one could become proficient at any given task through practice. Notwithstanding, there were times, psychologists added, when the neurological channels became unhealthy and resulted in the formation of pathological habits or obsessions. Athena Vrettos has usefully outlined these ideas:

> After the 1840s, theories of habit relied on the conservation of energy to explain the tendency of the mind to reinforce mental patterns, pathways, channels, or, to use Gaskell's suggestive term, "chains." While these patterns traced the flow of thought and were seen as the structural mechanisms of all learning, they could also induce a static form of development in which the individual rehearsed characteristic behaviors rather than evolving new ones. In later-nineteenth-century discussions of habit, this potential rigidification of human character appeared to pose a threat of psychological stasis that was often linked to deterministic forces of production and consumption in modern industrial society.[15]

Vrettos points to Miss Havisham's ghastly bridal paraphernalia as an example of such "rigidification" of human behavior. "Her decayed wedding gown, rotting banquet table, and Satis House," she writes, "have the same power to occupy the minds of others as Miss Havisham herself."[16] What is most interesting about Miss Havisham, in this context, is the way she routinely travels the same channels, mentally and literally, day after day. She walks round and round her room and lives the same moment continually in a way that demonstrates, to use Lewes's imagery, that she is caught in the same unhealthy channels. Whereas for some individuals, the neurological channels of physiological psychology were a strength, for Miss Havisham they have resulted in the development of crushing and debilitating obsessions.

In the latter half of the nineteenth century, a stream of texts emerged exploring how the sense of one's own physical failures could become a similar, unhealthy preoccupation. In 1864, for instance, Fyodor Dostoyevsky wrote *Notes from Underground* with the aim of showing how cultural pressures to be a certain way caused some people like his anonymous narrator to bury themselves in morbid solitude. The latter finds it impossible to interact with other people on a social level because of a belief in his awkward appearance:

> I very often looked at myself with frantic dislike, sometimes amounting to disgust, and therefore attributed the same attitude to everybody else. For example, I hated my face, I thought it was a scoundrelly face, and I even suspected there was something servile about it, and so every time I went to the office, I made agonizing efforts to seem as independent as possible, so that I should not be suspected of subservience, and to give my face the most well-bred expression I could manage.[17]

Interestingly, the narrator's perception of normality hinges on the idea of appearing "well-bred" and "noble." These terms link his painful sense of self-worthlessness with the notion of good breeding. His feelings that he will never display any of these attractions are exacerbated when he meets a young soldier in a bar:

> I had been standing by the table and unknowingly blocking the way; he wanted to get past, and took me by the shoulders and silently – with no warning or explanation – moved me from the place I stood to another . . . I had been treated like an insect. The officer was a six-footer; I was short and skinny.
>
> (52)

Clearly, the soldier's appearance is congruent with British ideals of masculinity: he is tall, strong, outgoing, and so on. After this meeting, the narrator "experience[s] an endless series of torments, crushing humiliations and attacks of spleen . . . I felt a spasmodic pain in my heart and hot flushes down my spine at the thought of the wretched inadequacy of my costume and the mean vulgarity of my small figure darting about" (55). He then buries himself in Miss Havisham-like solitude, cutting himself off from the influences that would probably cure his feelings of insignificance. This solitude is what he calls his "underground": "I missed life through decaying morally in a corner, not having sufficient means, losing the habit of living, and carefully cultivating my anger underground" (122). Unlike Esther Summerson and Jane Eyre, the narrator is not rescued from his morbidity by a female equivalent of Woodcourt and Rochester (although he comes close). Dostoyevsky's text is more pessimistic and doubly underscores the notion that certain methods of dealing with looks-related obsessions are uglier and unhealthier than the perceived flaws themselves. Psychologists would have argued that the narrator's insecurities become fixed ideas, channels (not shared by Jane and Esther) that drive the young man's actions to strange and extreme lengths.

In Wilkie Collins's *The Law and the Lady* (1875), Sarah Macallan also sees herself as a poor comparison with another character. She discovers from her husband's journal that the aptly named Miss Beauly is the true object of his affections. The discovery adds to Sarah's already insecure feelings about her appearance and she kills herself. In a suicide note addressed to her husband, she writes:

> [I am the] next worst thing . . . to a deformity – a plain woman. . . . I heard you once say (when you were not aware that I was within hearing) that the vainest women were always the ugly women. You might have accused me of vanity. Who knows?
> But I have no desire to shelter myself under this excuse. I am a jealous unhappy creature; always doubtful of your affection for me; always fearing that another woman has got my place in your heart.[18]

When her husband's journal confirms these fears, Sarah swallows fatal doses of arsenic which she had originally bought to improve her complexion:

> Yes: I have had enough of my life. Yes: I mean to die. . . . You don't know what it has cost me to control my hatred of [Miss. Beauly]. . . . The arsenic that I twice asked you to buy for me is in the dressing-case. I deceived you when I mentioned some common-place reasons for wanting it. My true reason was to try if I could not improve my ugly complexion – not from any vain feeling of mine: only to make myself look better and more loveable in your eyes. I have taken some of it for that purpose; but I have got plenty left to kill myself with. The poison will have its use at last. It might have failed to improve my complexion. It will not fail to relieve you of your ugly wife.
>
> (391–2)

Like Dostoyevsky's narrator, Sarah comes into contact with an individual who complies more with the nineteenth-century sense of what constitutes real beauty. This encounter, combined with her existing anxieties about being unattractive, plagues her with a sense of her own worthlessness. It is fitting that the poison bought to improve her looks actually becomes the means through which she kills herself. It is a device through which the novel highlights her death as the culminating point of a long-term disorder centered on her perceived ugliness. What Collins illustrates here is that Sarah's methods of dealing with her "unattractiveness" (poisoning her skin, then committing suicide) are, in fact, the ugliest things about her.

By 1891, then, psychiatry had a backlog of works, both literary and scientific, on which it could draw in order to identify and label the concept of a looks-related neurosis. In the spring of that year, Italian psychologist Enrico Morselli coined the term "dysmorphophobia." Derived from the Greek "*dysmorfia*," meaning "ugliness," dysmorphophobia meant, quite literally, the fear of being ugly. It is from this that the modern term "Body Dysmorphic Disorder" is taken. Differing little from modern definitions, Morselli described dysmorphophobia as "the sudden appearance and fixation in the consciousness of the idea of one's own deformity; the individual fears that he has become deformed or might become deformed, and experiences at this thought a feeling of inexpressible anxiety."[19] He continues:

> The ideas of being ugly . . . are not, in themselves morbid; in fact, they occur to many people in perfect mental health . . . . But, when one of these ideas occupies someone's attention repeatedly on the same day, and aggressively and persistently returns to monopolise his attention, refusing to remit by any conscious effort; and when in particular the emotion accompanying it becomes one of fear, distress, anxiety, and anguish, compelling the individual to modify his behavior and to act in a pre-determined and fixed way, then the psychological phenomena have gone beyond the bounds of normal, and may validly be considered to have entered the realm of psychopathology.
>
> (108)

Morselli drew upon the ideas of earlier psychiatrists like Esquirol. He toned down the idea of "insanity" to "psychopathology" but his indebtedness to earlier writers, both medical and literary, cannot be doubted. Morselli, like Freud, put a name to ideas that had long been in circulation. He confirmed, using a scientific vocabulary, what writers like Dostoyevsky and Collins had suspected for years: that people became obsessed by their perceived cosmetic failures.

In 1890, the year before Morselli coined "dysmorphophobia," Oscar Wilde wrote one of the century's most sustained literary analyses of the idea of human beauty. *The Picture of Dorian Gray* (1890, revised 1891) has had no shortage of critical interest. Yet, much is still to be gained by reading the novel in light of the developments that, by the 1890s, aimed to pinpoint and treat self-esteem-based pathologies. Unlike the characters discussed above, Dorian Gray does not believe himself to be ugly. Indeed, part of his obsessions stem from the fact that he is a remarkably beautiful man. The fears that haunt him are actually related to the prospect of losing his looks. Faced with his own portrait and Henry Wooton's hints that he will age, he realises the following:

> Yes, there would be a day when his face would be wrinkled and wizen, his eyes dim and colorless, the grace of his figure broken and deformed. The scarlet would pass away from his lips, and the gold steal from his hair. The life that was to make his soul would mar his body. He would become dreadful, hideous, and uncouth.
> As he thought of it, a sharp pang of pain struck through him like a knife, and made each delicate fibre of his nature quiver. His eyes deepened into amethyst, and across them came a mist of tears. He felt as if a hand of ice had been laid upon his heart.[20]

Notice, once again, how issues relating to a character's appearance begin to work a sickly influence over his thoughts. As a neurotic impression should, Dorian's realisation that he will age cuts through him like a knife. Indeed, that metaphor is not an unfitting one. As Gilman shows in *Creating Beauty to Cure the Soul* (1998), the late-nineteenth century was the period in which the concept and practice of cosmetic surgery was established.[21] In its frequent employment of knife imagery, Wilde's novel anticipates what would become the most popular method of dealing with perceived faults in one's appearance: plastic surgery. I will return to this point later.

The idea that Dorian develops dysmorphophobia is supported by the fact that his portrait gets uglier when he, in fact, does not. For Dorian, the portrait behaves like a mirror and, each time he looks at/into it, he sees an unattractive image (based on, though not truly resembling, his image):

> He drew the screen aside, and saw himself face to face. It was perfectly true. The portrait had altered.
>
> As he often remembered afterwards, and always with no small wonder, he found himself at first gazing at the portrait with a feeling

> of almost scientific interest. That such a change should have taken place was incredible to him. And yet it was a fact. Was there some subtle affinity between the chemical atoms, that shaped themselves, into form and color on the canvas, and the soul that was within him? Could it be that what that soul thought, they realized? – that what it dreamed, they made true? Or was there some other, more terrible reason? He shuddered, and felt afraid, and, going back to the couch, lay there, gazing at the picture in sickened horror.
>
> (93)

Dorian's behavior here echoes that of Esther Summerson. Like her, he melodramatically removes the screen in front of the "mirror" to see a face, which is seemingly not his own, looking back (it is worth noting that many individuals go through a similar experience following plastic surgery). Dorian is shocked by the hideous deformity represented in the portrait and wonders how the change has taken place. The one "terrible reason" that he does not consider in the above extract is that the portrait is actually unchanged but his ability to view it healthily is. In other words, it is possible that the portrait's ugliness is a fabrication of Dorian's dysmorphophobic mind. It is significant that the portrait is only ever seen by him after it begins to change. When his servants eventually find the mangled picture, it is exactly as its painter completed it: "They found hanging upon the wall a splendid portrait of their master as they had last seen him, in all the wonder of his exquisite youth and beauty" (213). Dorian himself, however, *has* changed: "Lying on the floor was a dead man, in evening dress, with a knife in his heart. He was withered, wrinkled, and loathsome of visage. It was not till they had examined the rings that they recognised who it was" (213). Dorian has become everything that he feared he would. His anxieties over his potential ugliness have seemingly worked their own fulfilment. It is not coincidental, I argue, that Dorian stabs the portrait and, in so doing, kills himself. The moment is narrated thus:

> He looked round, and saw the knife that had stabbed Basil Hallward. He had cleaned it many times, till there was no stain left upon it. It was bright, and glistened. . . . It would kill this monstrous soul-life, and without its hideous warnings, he would be at peace. He seized the thing, and stabbed the picture with it.
>
> (212)

Dorian's ideas about what his knife can accomplish echo those entertained by some individuals in relation to the scalpel. Observe how, like a scalpel, Dorian's knife cuts through one man, is cleaned, and then cuts through another. For dysmorphophobia sufferers, the scalpel, in cosmetic surgery, is often viewed as having the ability to "kill [a] monstrous soul-life" and restore the mind to "peace." Dorian's stabbing of the portrait is almost like a cosmetic procedure. He cuts into his own features as a way of dealing with the ugliness he sees written there. His drastic actions echo those of Sarah Macallan. She swallows arsenic as a means of improving her complexion but ultimately kills herself. Dorian, likewise, attempts to deal with *his* imaginary ugliness by lacerating his portrait. As with Sarah, this is an act of suicide. His final appearance as a "loathsome" corpse serves to underscore how his dysmorphophobic demons have finally won over the soul of the protagonist.

According to Gilman, the thing most valued by the dysmorphophobic individual was the prospect of invisibility. Writing about the advent of cosmetic surgery, he claims:

> (In)visibility is the goal of all aesthetic procedures. Aesthetic procedures are intended to move an individual from being visible in one cohort to being member of another cohort or collective, which is *so* visible that its visibility becomes defined as "normal." The "normal" defines itself as invisible. . . . It [is] important for the patient to become invisible.[22]

Gilman's ideas help us make sense of how we differentiate between vanity and dysmorphophobia (or BDD). Whereas the vain person wants to be conspicuous, the dysmorphophobic only wants to fit in. For the latter, there is a crushing need for their perceived flaws to be unnoticeable or, better still, gone.

Literary writings of the late-nineteenth and early-twentieth centuries, however, suggested that invisibility was a pathological and/or terrifying condition in itself. The resurgence of Gothic narratives at this time taught readers to fear the individual that could not be seen. Consider, for example, the Whitechapel murders of 1888. The most worrying aspect of Jack the Ripper was his apparent ability to evade detection and perform his gruesome work in silence. This is also a dominant fear in *Dracula* (1897). The Count's most uncanny trait is perhaps his ability to move ethereally from one form to another. It is also significant that he has no reflection. The dysmorphophobic individual, as written above, had a pathological fear of mirrors; the Count seeks to avoid them as well.

It is perhaps in H.G. Wells's *The Invisible Man* (1896), however, that the "horrors" of invisibility are most evident. When Griffin, the invisible man, first enters the narrative, he is wrapped in bandages. The other characters, villagers from "Bramblehurst," assume that he suffers from "a chronic irritation of the greatest intensity" and are prepared to meet with such a contingency.[23] As one character remarks:

> I tell 'e what I'm thinking. That marn's a piebald, Teddy. Black here and white there – in patches. And he's ashamed of it. He's a kind of half-breed, and the color's come off patchy instead of mixing. I've heard of such things before. And it's the common way with horses, as any one can see.
>
> (20)

It is assumed that Griffin is ashamed of his appearance when he is actually ashamed of the *lack* of one. It also becomes clear that he wears a false nose:

> The centre of his face became a black cavity. "Here," he said. He stepped forward and handed Mrs. Hall something which she, staring at his metamorphosed face, accepted automatically. Then, when she saw what it was, she screamed loudly, dropped it, and staggered back. The nose – it was the stranger's nose! pink and shining – rolled on the floor.
>
> (36)

False noses were seen as one of the "signs" of syphilis in the late-nineteenth century. The venereal condition, it was believed, frequently resulted in the sufferer losing his or her nose and being compelled to wear a false one.[24] "Artifical noses," writes Gilman, were "a sign that contemporary readers would understand as a visible mark of . . . disease."[25] Little wonder, then, that Mrs. Hall quickly drops that belonging to Griffin.

Yet it is not the invisible man's potential syphilis that disgusts Mrs. Hall but his way of hiding it. Before they understand that Griffin is invisible, the villagers resent him for hiding his condition and not meeting them frankly with the full extent of his unconventional appearance. As some of these characters remark:

> "Can't a man look at you? – Ugly!"
>
> And yet again: "Seemingly not. If the police was wanting you you couldn't be more wropped and bandaged" . . .
>
> "Looks a bit like a disguise, don't it? I'd like to see a man's face if I had him stopping at *my* place . . . ."
>
> (41)

The first line in this dialogue is playfully ambiguous. Does it, for instance, mean to call Griffin "Ugly" (as the capital letter would imply)? Or is the speaker suggesting that his inability to see the invisible man is itself an ugly situation? Either way, Griffin's method of dealing with his unconventional appearance is not well received by his fellow characters.

It is his actual invisibility, however, that is received with the most horror: "It was worse than anything. Mrs. Hall, standing open-mouthed made for the door of the house. Everyone began to move. They were prepared for scars, disfigurements, tangible horrors, but *nothing*!" (37, italics in original). These horrified reactions resonate with the aftermath of the Ripper murders. The most frightening type of assailant, it seems, is one that cannot be seen and, in Wells's novels, the invisible man does become antagonistic.

His invisibility, in fact, is but an exaggerated form of his previous attempts to hide himself away from public view. He explains to Mrs. Hall, early in the narrative:

> My eyes – are sometimes so weak and painful that I have to shut myself up in the dark for hours together. Lock myself up. Sometimes – now and then. Not at present, certainly. At such times the slightest disturbance, the entry of a stranger into the room, is a source of excruciating annoyance to me.
>
> (13)

Griffin's self-enforced solitude is like that of Dostoyevsky's narrator. Both characters have a desire to hide themselves from the prying eyes of the public and so, one might suggest, try to make themselves invisible (even without Griffin's experiments). Griffin is an albino, a condition that causes him to develop feelings of worthlessness. This is made apparent when he too comes into contact with military men: "I remember the barracks in Albany street, and the horse soldiers coming out, and at last I found myself sitting in the sunshine and feeling very ill and strange" (98). He is delighted, therefore, when he discovers how to make himself invisible:

> "I could be invisible!" I said, suddenly realizing what it meant to be an albino with such knowledge. It was overwhelming. I left the filtering I was doing, and went and stared out of the great window at the stars. "I could be invisible!" I repeated.
>
> (92)

As already noted, however, *The Invisible Man*, in confluence with a number of Gothic texts written around the same time, suggests that such invisibility is an ugly and pathological state.

It appears, therefore, that the dysmorphophobic individual was seen to be, more and more at the *fin de siècle*, an object of fear. His or her morbid predilection for suicide and invisibility was viewed as a worrying trait. Such impressions can only have been strengthened by the weighty influence of Mary Shelley's *Frankenstein* (1818). Victor's creature, although his ugliness is not imagined, becomes destructive when the human race teaches him to loathe his own image.

This is also the case with the last novel I wish to discuss here: Gaston Leroux's *The Phantom of the Opera* (1911). Like Griffin, Erik (the phantom) wears a false nose – a fact that has led Gilman to conclude that his characterisation draws on the Victorian spread of "syphilophobia."[26] The most interesting aspect of the phantom, however, is his method of dealing with his alleged unattractiveness. It is clear that this is a source of torture for Erik. Overcome by curiosity, the object of his unrequited affections, Christine Daaé, pulls off his mask. She narrates:

> Oh, horror, horror, horror! . . . If I lived to be a hundred, I should always hear the superhuman cry of grief and rage which he uttered when the terrible sight appeared before my eyes . . .
> "Look! [Erik cries] You want to see! See! Feast your eyes, glut your soul on my cursed ugliness! . . . Well, are you satisfied? I'm a very good-looking fellow, eh?" . . .
> And he seized my hands and dug them into his awful face. He tore his flesh with my nails, tore his terrible dead flesh with my nails![27]

Erik then expresses "every emotion, every suffering of which mankind is capable of suffering" (129). The anger he feels at Christine seeing his face is vented in violence against himself. Unable to understand why Christine might pity him, he forces her to inflict the violence he assumes he deserves. Like Dostoyevsky's narrator, Erik also buries himself underground – beneath the opera house. Despite beginning with self-flagellation, his neuroses then develop in such a way that he becomes a danger to others. It is at these times that his ability to become invisible seems most terrible. Moncharmin and Richard, the opera managers, for instance, hear the phantom's voice without being able to detect its whereabouts:

> They had felt his breath. Moncharmin's hair stood on end. Richard wiped the perspiration from his forehead. Yes, the ghost was there,

> around them, behind them, beside them; they felt his presence without seeing him, they heard his breath, close, close, close to them!
>
> (79)

That Erik is able to instil such terror with his trickery is a good indication of how Leroux draws on a range of Gothic narratives which suggested that any man's self-created invisibility is a worrying characteristic. Erik kills several people and tries to blow up the opera house. Yet, even so, the narrative closes with a pitying glance of what he might have been had it not been for his ugliness:

> Poor, unhappy Erik! Shall we pity him? Shall we curse him? He asked only to be "someone," like everybody else. But he was too ugly! And he had to hide his genius *or use it to play tricks with*, when, with an ordinary face, he would have been one of the most distinguished of mankind! He had a heart that could have held the empire of the world; and, in the end, he had to content himself with a cellar. Ah, yes, we must needs pity the Opera ghost.
>
> (264, italics in original)

Erik's wish to be "like everybody else" gels with the dysmorphophobic desire to be inconspicuous. That Erik's hideousness is real (and not imagined) is not as important as the way in which he chooses to deal with it. He allows his obsessions to drive him underground and he resorts to murder. *The Phantom of the Opera* fits in with a series of writings from the *fin de siècle* which sought to identify some individual reactions to ugliness as pathological, worrying, and even dangerous.

In conclusion, the nineteenth century's response to the concept of degeneration has often been vilified as prescriptive, ideological, and extreme. The notion of ethnic cleansing, for instance, is one of the many Victorian ideas written about with a view to demonstrating how that period sought to condition itself in a way that reinforced established delegations of power. In this chapter I have intended to show how the period's reactions to procreative pressures were more complex and varied than is often assumed. While I do not doubt that theories of healthy breeding had subjugating cultural effects, writers like Dickens, Morselli, and Wilde also demonstrated how such pressures led to the formation of image-based obsessions. In texts like Dostoyevsky's, Wells's, and Leroux's, moreover, readers were presented with the idea that individuals who did develop such obsessions could become more worrying still. While warnings against the perceived decline of humanity abounded at the *fin de siècle*, so too, it seems, did narratives showing the potentially devastating personal effects of such ideas.

Notes

I would like to thank Anne Stiles, Joseph Bristow, and members of the British Association for Victorian Studies for their helpful suggestions.

1. Katharine A. Phillips, *The Broken Mirror: Understanding and Treating Body Dysmorphic Disorder* (1986; Oxford: Oxford University Press, 1998), 3–4. Italics in original.
2. Francis Galton, "Hereditary Talent and Character: Part I," *Macmillan's Magazine* 12 (1865), 157–66, 165.
3. See [Eneas Sweetland Dallas], "On Physiognomy" and "The First Principle of Physiognomy," *Cornhill* 4 (1861), 472–81 and 569–81.
4. See especially Sally Shuttleworth, *Charlotte Brontë and Victorian Psychology* (Cambridge: Cambridge University Press, 1996), 57–70 and Lucy Hartley's *Physiognomy and the Meaning of Expression in Nineteenth-Century Culture* (Cambridge: Cambridge University Press, 2001).
5. See Andrew Smith, *Victorian Demons: Medicine, Masculinity and the Gothic at the Fin-de-Siècle* (Manchester: Manchester University Press, 2004), 95–117.
6. Sander L. Gilman, *Creating Beauty to Cure the Soul: Race and Psychology in the Shaping of Aesthetic Surgery* (Durham: Duke University Press, 1998), 34.
7. Shuttleworth, *Charlotte Brontë and Victorian Psychology*, 82.
8. Jane Wood, *Passion and Pathology in Victorian Fiction* (Oxford: Oxford University Press, 2001), 64.
9. Charlotte Brontë, *Jane Eyre*, ed. Margaret Smith and Sally Shuttleworth (Oxford: Oxford University Press, 2000), 98. Subsequent page references appear in parenthesis.
10. Charles Dickens, *Bleak House*, ed. Andrew Sanders (London: J.M. Dent, 1994), 218. Subsequent page references appear in parenthesis.
11. See Phillips, *The Broken Mirror*, 28–9.
12. Jean Étienne Dominique Esquirol, *Des Maladies Mentales, Considérées sous les Rapports Médical, Hygiénique et Médico-légal* (1838); trans. E.K. Hunt as *Mental Maladies: A Treatise on Insanity* (Philadelphia: Lea and Blanchard, 1845), 28.
13. G.H. Lewes, *Problems of Life and Mind* (1874–9), partly reprinted in Jenny Bourne Taylor and Sally Shuttleworth, eds, *Embodied Selves: An Anthology of Psychological Texts 1830–1890* (Oxford: Oxford University Press, 1998), 89–91, 89.
14. See the introduction to this volume for more information on the relationship between psychology and neurology.
15. Athena Vrettos, "Defining Habits: Dickens and the Psychology of Repetition," *Victorian Studies* 42 (Spring 1999/2000), 399–426, 400.
16. Vrettos, "Defining Habits," 411.
17. Fyodor Dostoyevsky, *Notes from Underground* and *The Double*, ed. Jessie Coulson (London: Penguin Books, 2003), 47. Subsequent page references appear in parenthesis.

18. Wilkie Collins, *The Law and the Lady*, ed. Jenny Bourne Taylor (Oxford: Oxford World Classics, 1992), 388. Subsequent page references appear in parenthesis.
19. Enrico Morselli, "Dysmorphophobia and Taphephobia: Two Hitherto Undescribed forms of Insanity with Fixed Ideas," ed. Laurence Jerome, *History of Psychiatry* 12 (2001), 103–14, 107. "Taphephobia" is defined by Morselli as the fear of being buried alive – a fear that, he says, reached obsessive proportions in many Victorians.
20. Oscar Wilde, *The Picture of Dorian Gray,* ed. Robert Mighall (London: Penguin, 2000), 27. Subsequent page references appear in parenthesis.
21. Gilman, *Creating Beauty*, 21.
22. Gilman, *Creating Beauty*, 4.
23. H.G. Wells, *The Invisible Man*, ed. Patrick Parrinder (London: Penguin, 2005), 20. Subsequent page references appear in parenthesis.
24. That is before surgery was capable of constructing a new nose.
25. Gilman, *Creating Beauty*, 33.
26. Gilman, *Creating Beauty*, 31–8.
27. Gaston Leroux, *The Phantom of the Opera* (London: Penguin Popular Classics, 1995), 127–9. Italics in original. Subsequent page references appear in parenthesis.

4

Doctor Zay and Dr. Mitchell: Elizabeth Stuart Phelps's Feminist Response to Mainstream Neurology

Kristine Swenson

Made internationally famous by the publication of her post–Civil War spiritualist work, *The Gates Ajar* (1868)—a novel outsold in the nineteenth century only by *Uncle Tom's Cabin*—Elizabeth Stuart Phelps became in the last decades of the century a leading figure in American literature.[1] Phelps used her prominence to advocate for spiritual uplift and social change—especially "Heaven, homeopathy, and women's rights."[2] As a self-described "professional invalid" who suffered from a chronic "nervous disease,"[3] Phelps interested herself in the medical education and practice of women, befriending and consulting several women physicians, and in her writing "advanced compelling critiques of the medical profession and its treatment of women."[4] In, for instance, two pieces written in 1874, Phelps disputes the claims of Harvard Medical School Professor E.H. Clarke that higher education causes physiological harm to women.[5] A decade later, Phelps initiated a correspondence with the eminent neurologist Silas Weir Mitchell in which they discussed their mutual interest in medical fiction, and Phelps challenged Mitchell by advocating for women physicians and a homeopathic treatment of neurological illnesses. Phelps's feminist and homeopathic response to Mitchell and mainstream neurology is the subject of her 1882 novel, *Doctor Zay*, in which a female homeopath cures and simultaneously reforms a male neurasthenic.

As her correspondence with Mitchell attests, Phelps breaks with mainstream neurology and medical ideology on the questions of the appropriate treatment of the male neurasthenic and the desired result.[6] Using the more "heroic" medicine that Phelps and homeopaths called "allopathy," Mitchell would have treated men and women differently and for different results. Orthodox medical treatments for neurasthenia were deeply gendered, replicating Victorian stereotypes of manly men and womanly women and reinforcing the idea that the "highly developed differentiation of the sexes . . . was both a sign of and a necessity for civilization."[7] Thus, while nerve specialists such as Mitchell prescribed the now-infamous "rest cure" for delicate, nerve-exhausted women, they advised neurasthenic men, after a period of rest, to "go West" or otherwise seek out physical challenges that would strengthen their natural constitutions.[8] Phelps's homeopathic Doctor Zay, in contrast, treats male and female patients similarly with gently stimulating remedies and toward a similar, hybridized result.

Homeopathy gained widespread popularity in the U.S. during the 19th century, especially among social progressives and spiritualists such as Phelps, because of its gentler and less invasive approach to curing illness, its emphasis upon the body's natural ability to cure itself if stimulated correctly, and because of its greater acceptance of women practitioners.[9] Central to homeopathic medicine is the law of *simila* or "let like cure like,"[10] whereby practitioners prescribe small doses of medicines that create "in the body an artificial form of the same illness, but one which the body [is] able to combat."[11] To Phelps, neurology and homeopathy were a natural pairing since her own illness was, she believed, purely "nervous" and doctors seemed to her "rather psychologists than physiologists."[12] With its holistic approach to an integrated body and mind and emphasis upon the spiritual dimension to healing practices, homeopathy allowed Phelps, alongside other novelists of her period, to critique the materialism of mainstream neurology.[13]

Homeopathy's emphases upon the joint work of the mind and spirit with the body, and upon the need for mental calm in order for the body to cure itself, sound much like many of Mitchell's treatments for nervous exhaustion, particularly in women, the more sensitive and spiritual sex.[14] The potentially threatening affinity between his own new methods and those of "quack" doctors perhaps explains some of Mitchell's contempt for homeopathic medicine as well as for women doctors, who were generally viewed as more nurturing and sympathetic than men.[15] Little has been written on specifically homeopathic neurology, perhaps because homeopathy, along with other sectarian medical practices, was waning in influence even as neurology was being solidified and legitimated. Homeopathic *treatment* of neurological disorders, as Phelps's own experience attests, continued through the nineteenth century. While Mitchell treats homeopathy as an illegitimate medical aberration, Phelps sees it as a feminist alternative to Mitchell's sort of medicine.

Phelps's alignment of allopathy with male medicine and homeopathy with female medicine is not strictly accurate in historical terms. Though

Lilian R. Furst asserts that homeopathy, with its small and gentle medicinal doses, was "more favored by women physicians than the heroics of allopathy,"[16] Regina Morantz-Sanchez would argue that, in fact, most women physicians felt great pressure to conform to medical orthodoxy. While a greater percentage of women than men called themselves "homeopaths" (or practiced other forms of sectarian medicine), the overwhelming majority of women physicians, in fact, practiced orthodox medicine.[17] Nevertheless, Phelps's use of the female homeopath to cure ills of gender as well as those of mind and body is a neat way to lay out her feminist agenda regarding "ideal marriage" in which the New Man and the New Woman are remarkably similar, complementary figures.

Of all her novels, Phelps's *Doctor Zay* explores most explicitly and optimistically how a woman might balance love and a profession, a question that became central to New Woman fiction of the 1880s and 1890s. Sarah Grand coined the phrase "new woman" in 1894 to describe the modern woman who has rejected the "Home-is-the-Woman's-Sphere" ideology.[18] Grand, however, makes it clear that the New Woman is not, in fact, new: "the new woman is a little above [conventional man], and he never even thought of looking up to where she has been sitting apart in silent contemplation all these years. . . ."[19] Moreover, though the New Man was not named until 1911,[20] his presence as the necessary counterpart to the New Woman was assumed much earlier, as Grand's essay—whose subject is actually more manhood than womanhood—shows: "The man of the future will be better, while the woman will be stronger and wiser."[21]

Though writing a decade earlier than Grand, Phelps engages the same questions of gender roles and sexual politics of later fiction in strikingly similar terms. Zay tells her would-be suitor, Yorke, "you have been so unfortunate as to become interested in a new kind of woman. The trouble is that a happy marriage with such a woman demands a new type of man" (244). Again, as does Grand in her novels *Heavenly Twins* (1893) and *The Beth Book* (1897), Phelps uses medical tropes to help conceptualize her new woman and man. Whereas Grand asserts masculine degeneration through a sexually diseased and corrupt male doctor, Phelps afflicts Yorke with the less sensationalistic disease of male neurasthenia and offers him a gentle, homeopathic cure at the hands of the honorable Dr. Zay. Phelps's characterization of Yorke's illness, though, is not merely a figurative way to explore gender. Jane Wood notes that "in fiction and in medicine, the configuration of the male nervous invalid was shaped and informed by the changing perceptions of manliness" and that by "exploring the contradictions and ambiguities within medicalized definitions of gender, representations of invalidism in literary texts challenge the defining categories of disease classification."[22] Nor is Phelps's choice of profession for her New Woman merely figurative. Zay's position as a homeopathic M.D. is both historically plausible and thematically rich. Though early women physicians faced tremendous obstacles to professional success, medicine was one of the few professions open to women in the 1880s; by 1900, women "comprised close to five percent of the profession, over seven thousand strong."[23] In the United States especially, the figure of the woman doctor enjoyed a sort of vogue during the last decades of the century, and was closely associated with reform movements including critiques of "the impersonal, dehumanized standards of scientific medicine."[24] At the same time, Zay's profession allows Phelps to intensify traditional character development in her novel by portraying psychological and physiological change in her characters in medically informed and specific ways.

When a Boston lawyer, Waldo Yorke, drives his buggy off a bridge in rural Maine, the local homeopath, Dr. Atalanta (Zay) Lloyd, brings him back to life and health. During his illness, Yorke falls in love with Zay, who, however, resists the advances of her patient until his nerves are cured and his character has been transformed. Like the Atalanta of classical myth, Zay eludes marriage until she is ultimately snared and won over by Yorke. *Doctor Zay* has generally been read as one of a group of woman doctor novels, a subgenre that became something of a fad in the United States and Britain in the 1880s and 1890s.[25] Writers such as Henry James, Sarah Orne Jewett, Charles Reade, Arabella Kenealy, and many others explored questions of the modern woman's education, independence, and sexuality through the figure of the woman doctor, a representative New Woman who, by her very being, challenged orthodox medical opinion about woman's biology and intellectual ability. In this reading, *Doctor Zay* is about the eponymous heroine, an early example of American New Womanhood—or, as Phelps would say, ideal womanhood—who struggles to balance her career with a desire for love and marriage. Among woman doctor novels, *Doctor Zay* is remarkable for the strength and stability of its heroine—for the fact that she retains her career and professionalism without compromises and wins marital love on her own terms.

Indeed, *Doctor Zay* is often read as a direct response to W.D. Howells's *Dr. Breen's Practice*, published a year earlier and also in the *Atlantic Monthly*. In contrast to Phelps's medical woman of integrity, talent, and strong nerves, Howells's Dr. Breen is a "thing of hysterical impulses" who is "too nervous" to be a doctor.[26] Though a "strong supporter of women's suffrage,"[27] Howells, nevertheless, maintains the underlying assumption about woman's dangerously sensitive nature that was the cornerstone of theories such as Clarke's, Mitchell's, and other medical men of the period. *Doctor Zay* was published in the *Atlantic* with a letter from Howells explaining that the similarities between their two novels were coincidental and that he had, as "Editor of The Atlantic," encouraged her to write and submit her novel for publication.[28]

In truth, the similarities between the two novels are unlikely to be coincidental; the evidence suggests that, through *Doctor Zay*, Phelps takes issue with Howells's sexist claims about women and medicine just as she had Clarke's and would Mitchell's. For one, Phelps did not write to Howells about *Doctor Zay* until after the first installment of *Dr. Breen's Practice* had appeared in the *Atlantic* in August of 1881.[29] Equally significant is that Phelps had been publishing essays and short fiction advocating for women doctors since the early 1870s, before Howells had shown any interest in the topic. "In none of her contemporaries," writes Frederick Wegener, "does one find the kind of sustained and pervasive engagement with the figure of the woman doctor reflected so abundantly in Phelps's work up to the early 1880s."[30] Responding to Weir Mitchell's criticism of *Doctor Zay*'s credibility, Phelps writes, "Were I an old friend . . . I should take you to task a little for what you say of women physicians . . . I *know* women physicians thoroughly . . . Every fact and figure that I gave in Doctor Zay was *understated*, from what I know to be the existing realities."[31]

Phelps felt protective of the woman doctor, and although she certainly admired Howells's skill as a novelist, she could not have been pleased by his characterization of Dr. Breen, who takes up medicine when thwarted in love and then abandons her profession when an appropriate marriage offer comes along. Similarly, although Phelps respected Mitchell's medical skill, she openly defended to him her support of women doctors and homeopathy, two modern medical innovations that he deplored. *Doctor Zay*, then, uses a homeopathic neurology to oppose conservative medical and cultural practices by championing the radical figure of the female physician, by challenging gender roles for both women and men, and questioning the relative merits of marriage and career for women. When Zay, using "female" homeopathy, cures the male neurasthenic and then transforms him into an acceptable husband, she provides a feminist alternative to both mainstream allopathic neurology and traditional gender practices.

Phelps sought an ending for her New Woman doctor that was quite different from that of Howells's Grace Breen. Other novelists found it simpler to write successful woman doctors as women who remain single.[32] Sara Orne Jewett's Nan Prince from *A Country Doctor* is, for instance, an instructive contrast to Zay and arguably a more believable character. Phelps, however, wanted to demonstrate that marriage and a career were not irreconcilable for women; she wanted, in fact, to represent the reality of professional women as she knew it. Having investigated "medical schools where women collect," Phelps found that roughly half of women medical students were married: "It does not matter how they *can* [pursue medical careers while married]; it is enough that they *do*!"[33] In the interests of realism, of expanding professional opportunities for married women, and—frankly—of appealing to readers who preferred romantic closure, Doctor Zay should marry. However, as Chris Willis notes, writers who wished to portray their New Women triumphing over social obstacles faced a difficult problem: for "the nineteenth-century heroine, closure almost inevitably meant marriage. But how could the New Woman be found a suitable mate?"[34]

Writers as politically diverse as George Gissing, Olive Schreiner, and Charlotte Perkins Gilman all sought to create an appropriate match for the New Woman; most often, however, their New Men are unsatisfying as lovers, underdeveloped as characters, or both. These New Men tend to be sophisticated enough to be interested in New Women but almost inevitably fail them unless the women bend to their will. Often, the men turn out to be not as enlightened as they had claimed or believed themselves to be. Phelps seems well aware of this problem with New Men and yet desires both to maintain the integrity of her heroine and to achieve romantic closure. Phelps's solution to the problem of the New Man is for the New Woman heroine in *Doctor Zay* to constitute her own ideal mate using her homeopathic medical training.

Despite its title, the focus of *Doctor Zay* is much more the psychological evolution of the doctor's patient and suitor, Waldo Yorke, than about the doctor herself. The novel is told from Yorke's perspective, and it is his, not Zay's, nervous system that is under scrutiny. Phelps acknowledges as much in a letter to Weir Mitchell: "I have [in *Doctor Zay*] tried to draw a Doctor by reflection, or by reflex action; the result is, at least I hope, a patient."[35] Phelps was herself a long-term sufferer of the sorts of nervous maladies that she inflicts upon Yorke, and by the time she wrote *Doctor Zay* she had undergone treatment for her illness at the hands of male and female physicians, allopaths, and homeopaths. Her portrait of the neurasthenic and of his medical treatment thus contains realism and depth of sympathetic identification. As a result and in contrast to the majority of New Men in fiction, the portrait of Waldo Yorke in *Doctor Zay* is remarkably well developed and psychologically believable.

Even more remarkable, what Phelps shows us is not a New Man already formed whose depths must be drawn out by the ideal woman, but the actual sexing of the New Man through illness and feminization at the hands of the woman doctor. By making her New Woman a physician (rather than, say, a writer or artist), Phelps foregrounds Victorian assumptions about sex and gender that mainstream physicians such as Mitchell believed to define the natures—and certainly the nervous systems—of men and women. Because she is a physician, an expert in "psychological investigation" with access to the human heart even greater than a priest's, Zay can diagnose and treat Yorke's nervous unmanning and articulate this process for the reader.[36] In this way, we gain access not only to what one Victorian feminist thought the New Man might *be* but how he might *become*.

Phelps's formula for the New Man begins with a fundamentally sensitive, educated, and cultured man whose mother has already exposed him to "advanced ideas." The problem with such a man, with Waldo Yorke at the novel's beginning, is that he tends to be at once socially superfluous—neither useful, strong, nor ambitious—and yet retains many of the traditional prejudices and the selfishness of the privileged male. Yorke is a "distinguished-looking" but "nervous" bachelor of twenty-eight who both yearns for and sneers at the concept of the ideal woman (7, 9). Yorke diagnoses his own character for Doctor Zay's benefit: "Inherited inertia. Succumbed to his environment. Corrosion of [Boston's] Beacon Street. . . . Native indolence, developed by acquired habit. Hopeless correlation of predestined forces. Atrophied ambition. Paralyzed aspiration. No struggle for existence" (166). While not disagreeing with this diagnosis, Zay softens and complicates it, describing Yorke's character in the most current medical terminology of the day: "I should rather call it [a case] of hyperaesthesia. . . . Superfluous, and therefore injurious, sensitiveness. You experience a certain scorn of the best into which you know yourself capable of resulting" (167–8). By diagnosing hyperaesthesia—or "excessive and morbid sensitiveness of the nerves or nerve-centres"[37]—Zay has linked Yorke firmly to the neurasthenic and hysteric. And Yorke describes himself as Weir Mitchell would have the male neurasthenic. In "Wear and Tear or Hints for the Overworked" (1871), Mitchell explores how upper class men of New England, in particular, have been "wastefully spending" the "capital of vitality" built up within the race by their physically hard-working ancestors. Contributing to the great evil of nervous exhaustion, claimed Mitchell, were:

> the cruel competition for the dollar . . . the racing speed which the telegraph and railway have introduced into commercial life, the new

> value which great fortunes have come to possess as means towards social advancement, and the overeducation and overstraining of our young people.[38]

Yorke, in short, is an example of the "over-civilized" urban man that worried so many American commentators at the turn of the century when various nervous disorders (often lumped into the catch-all category of neurasthenia) seemed to threaten the vitality of the middle and upper classes.[39] Paradoxically, many commentators, including Weir Mitchell, believed the high rates of neurasthenia in America to be a sign of the nation's evolved "cultural superiority."[40]

With his nervous sensitivity, Yorke represents a type of man believed to be increasingly common in this period. An obsession with male vitality developed in the last decades of the nineteenth century in Europe and the U.S. as a cultural panic about evolutionary "fitness" altered how Victorians thought about sex. The task of sexologists and physicians, argues Angus McLaren, shifted dramatically in the 1880s and 1890s from a preoccupation with controlling excess (nymphomania and satyriasis) to avoiding enervation: "Their concern, faced with declining birth rates that they attributed in part to the modern plagues of frigidity and impotency spread by an overcivilized society, was with the new problem of how to guarantee 'arousal.'"[41] Eugenicists, McLaren adds, heightened the panic by suggesting that the "quality as well as the quantity of future generations" was threatened.[42] The novel's awareness of this post-Darwinian panic about manhood is shown by Yorke's own self-diagnosis, replete as it is with the language of degeneration and especially with diction that emphasizes Yorke's lack of vitality and energy. Sex and romance thus become entangled with neurology in Yorke's illness and cure, an entanglement that will in later decades be cast as explicitly eugenic in its implications.[43] Though in Phelps's novel the impact of marriage upon the woman is the main concern, in novels by Sarah Grand, Charlotte Perkins Gilman, and Arabella Kenealy, it is clear that no less than the future of "the race" is at stake in the mating of the New Woman and Man.

Phelps seems to concur with Mitchell and other orthodox physicians about the manly deficiencies of men such as Waldo Yorke.[44] Yorke finds himself in Doctor Zay's hands in the first place because of his own idle recklessness. A licensed lawyer, Yorke retains no clients and so is free to travel to the backwoods of Maine to investigate for himself a minor inheritance. "You are not a great man," Zay tells Yorke frankly, "but you have uncommon qualities.... You could be successful if you chose" (169). Doctor Zay understands intuitively that the neurasthenic Yorke, though capable of expressing ardent love for an advanced woman, will ultimately become dissatisfied with her. A "happy marriage with [an ideal] woman demands a new type of man" she tells him, "By and by you would chafe under this transitional position. . . . You would think of the other men, whose wives were always punctual at dinner in long dresses. . . . You would not know how to cultivate happiness with a woman who had diverged from her hereditary type" (244).[45]

Although he desires the ideal New Woman, Yorke is in no fit state to be satisfied by her. But Mitchell's heroic type of cure that insisted upon sharply differentiated sex roles would make Yorke even less satisfied with a New Woman wife. An altogether different sort of cure is thus required if Yorke is to become the New Man who can happily marry Dr. Zay. To become Phelps's "new type of man," Yorke must endure prolonged physical and emotional pain, an unmanning invalidism, and a course of homeopathic treatment of body and character at the hands of his female physician that ultimately makes him both strong enough and sympathetic enough to be part of an ideal marriage. He must, in short, be feminized and then remasculinized by a strong woman.

From their first meeting, the traditional gender roles of the doctor and patient are reversed. Zay has an active medical practice that takes her out at all hours; she is strong, independent, and competent. She is also difficult to read, for we never get inside her head. In contrast, Yorke lies on his invalid couch, dependent upon Zay for his well-being and awaiting her return. He spends much of his time speculating about her, allowing himself to fall in love with her, and so disturbing his already sensitive "nerves." In the early days of his treatment, Yorke exhibits excessively feminine behaviors: he grows "giddy and faint" at the sight of his wounded shoulder, and faints "quite away" while watching Zay stitch it up (46, 48); he begs his physician to stay with him, is "restless" waiting for her return, and attends to his own "sensations anxiously" (58).

As Yorke gains physical strength, both he and his doctor place more emphasis upon the nervous strain that his accident has caused. Zay prescribes for Yorke standard homeopathic remedies—aconite pellets for shock and restlessness[46] when he would prefer "to be talked to, as if she had been like other women" (59). Yorke's pallor and burning dark eyes do not seem romantic to Zay who, notes the narrator, "was not an artist" (61). Rather, she says to herself, "'That is a *cinchona* look'" (61), cinchona being a standard homeopathic treatment for digestive ailments as well as headaches and exhaustion.[47] Following the publication of John Erichsen's *On Railway and Other Injuries of the Nervous System* (1866) and then George Beard's *American Nervousness* (1869), doctors believed that after an accident such as Yorke's that involved "spinal concussion," conditions "such as spinal irritability and neurasthenia, or even hysteria and malingering, could occur."[48] Zay would have anticipated and understood Yorke's neurasthenic symptoms as the logical result of his accident.

In this phase of recovery, Yorke, too, is fully conscious of his feminized state, often commenting on it ironically and occasionally expressing his new sympathetic identification with women who, more often than men, find themselves invalids or otherwise dependent upon others. Yorke chafes under his "unprecedented consciousness of dependence" and, after a day that for him "dragged piteously," exclaimed to himself, "'How *dare* men ridicule or neglect sick women?'" (84, 99). Yorke's enforced role reversal and this identification with women are vital to his development as a New Man. To be Zay's match, he must not only understand rationally but also *feel* the need of the women Doctor Zay serves, the "refinement and suffering and waste of delicate life" that only the "woman healer" can address (75).

But his situation and newfound feeling also cause him to fall in love with Zay in a sort of Freudian transference that male doctors regularly expected from their female patients. Though we understand later that Zay, too, is falling in love during her treatment of Yorke, she nevertheless maintains her professionalism and treats his behavior as symptomatic of his illness: "You are not in love," Zay tells her patient, "you are only nervous" (191). "'You consider my love a symptom,'" Yorke accuses her, "'Because I am a patient, am I not a man? . . . am I to be set aside like a hysteric girl, for the state of whose limp emotions her medical attendant feels in honor bound to look out?'" (197, 198). While Yorke is in a "feminine" state of physical disability and nervousness, Zay feels no danger from him and so treats him with kindness and diligent attention; she treats him, in short, as she would one of her helpless women patients. But as he regains strength and health, she distances herself from him and works to strengthen rather than indulge his nerves.

At this point in the novel and in Yorke's recovery, it becomes clear that Zay's agenda (and Phelps's) is to work on the underlying character of the man as much as the illness that has exacerbated and revealed that character. Such holistic attention to the patient is fundamental to homeopathic medicine, which Phelps also speaks of as closer to psychology than physiology.[49] Instead of large doses of medicines favored by allopaths, Zay prescribes natural remedies—chamomile, aconite, and carbo—massage, and the occasional tablespoon of brandy, all designed to gently push Yorke's symptoms—morbid sensitiveness to trifles and his continued pursuit of her—in order to encourage the body to mend itself.[50]

Through much of the novel, this strategy of Zay's appears purely clinical. She has diagnosed her patient as hyperaesthetic and treats his symptoms. Yet before long it becomes clear that part of his cure—not just one of his symptoms—involves stimulating contact with *his* ideal "similar," Zay. Romance thus becomes entangled with medicine in Yorke's cure. When Yorke dares to flirt openly with Zay, she replies—after a silence "fraught with significance"—"I believe I shall give you Nux," causing him to flush "painfully" (81). Apparently, Yorke knows his homeopathic remedies well enough to understand this half-playful reproof, for nux vomica would have been most often prescribed for "extreme over-sensitivity . . . irritability," and gastric upset particularly as a result of "over-indulgence."[51]

Although Zay gives Yorke various homeopathic tonics, the best medicine seems to be what he drinks in from her presence. He gains strength—and even manliness—from her; or, in homeopathic terms, the small bit of masculine energy in her stimulates his own system's production of manly power. Yorke notes that Zay "possessed a power that was far more masculine than feminine," and he absorbs this power "idly, as he did returning life and the throbbing day" (131, 133). Her "cool, scientific eye . . . stimulated more than it defied him" (133). And when she asks for his help with a patient, he feels suddenly like her, "excited and strong" (144). Yorke credits Zay with instilling in him both this newfound masculine strength and his earlier feminine sympathy. As do many writers and physicians in the nineteenth century, Zay and Yorke use evolutionary language to discuss both sexual politics and progress from illness to health. However, Phelps once again is not shy about using such language in unorthodox ways toward feminist or idealist ends. Even as he is rediscovering his masculinity (in its New form), Yorke comments amusedly that he "shall make rather a superior woman by the time I get well" with "these beautiful characteristics that suffering or you [Zay]—it's all the same thing . . . —[a]re cultivating in me" (134). He notes, however, that he is "*not* a woman . . . only an arrested development" and that this "host of feminine virtues" that she has instilled in him won't live long (134).

Interestingly, Zay suggests that Yorke should consider this period of superior womanhood as "a rudiment," a comment that relates directly to Phelps's attitude to the science of her day and to how we should read her "sexing" of Yorke (134). In both *Origin of Species* and *The Descent of Man*, "rudiment" means something like "vestigial"—organs suppressed through natural selection or "disuse" that although "liable to occasional reappearance through reversion" are "absolutely useless, such as the mammee of male quadrupeds."[52] Indeed, for Darwin the presence of rudimentary organs is strong proof of sexual differentiation between men and women. Writers in the broader culture, however, maintained the older, less specialized meaning of the word "rudiment" even as they nodded to its evolutionary implications. A pre-Darwinian use of this sort occurs in a letter written by Geraldine Jewsbury to Jane Welsh Carlyle in 1849: "There are women to come after us, who will approach nearer the fulness of the measure of the stature of a woman's nature. I regard myself as a mere faint indication, a rudiment of the idea, of certain higher qualities and possibilities that lie in women."[53] If—as I think unlikely, despite the evolutionary bent of her conversation—Zay is using "rudiment" in strict Darwinian terms, then Yorke's host of feminine virtues are, as he suggests, temporary and useless. The context of the novel, however, suggests that Zay's use of the term is closer to Jewsbury's; she is implying that Yorke will not lose or again suppress those feminine virtues but build upon them toward a higher evolutionary state.

Mainstream neurologists, including Weir Mitchell, shared Darwin's assumption that the basic differences between men and women resulted from long evolutionary development and were therefore not amenable to quick alteration: "[T]he grave significance of sexual difference controls the whole question [of nervous exhaustion]," writes Mitchell, "The woman's desire to be on a level of competition with man and to assume his duties is, I am sure, making mischief . . . [N]o length of generations of change in her education and modes of activity will ever really alter her characteristics. She is physiologically other than the man."[54] Not only does this statement explain Mitchell's response to women practicing medicine, it also suggests that Mitchell's reading of Yorke's "feminine" virtues would have differed fundamentally from Zay's. He would, in fact, have been likely to read them as rudimentary in Darwin's sense of the word and as Yorke himself would like to believe them—as a temporary reversion to a more primitive form brought on by the utter exhaustion of Yorke's manly nervous system.

Like many other post-Darwinian feminists, Phelps altered the evolutionary hierarchy established by Darwinists and accepted by orthodox medical science in order to argue for woman's position of moral superiority over men. The evolutionary hierarchy that Zay assumes and that Yorke accepts with some amused detachment places "old" women above "old" men and, above these, New (advanced, superior, ideal) Women, who are superior to New Men. Both sexes must have reached this higher level to achieve what Zay terms an "absolute marriage," a rare partnership with an "element of twinship" (160). Again, Phelps's ideal partnership of man and woman, which seems to invoke the homeopathic law of similars, is diametrically opposed to that of Mitchell and most other male physicians and evolutionary theorists of the day, all of whom assumed that highly differentiated sexes were evolutionarily natural and culturally desirable.

Phelps had also been playing with gender reversal in medical situations before writing *Doctor Zay*. "Zerviah Hope," a story about a yellow fever epidemic set in South Carolina, features Dr. Marian Dare, a noble woman doctor similar to Zay, and Zerviah Hope, who travels south to volunteer to nurse the dying.[55] In many respects, this medical story is quite different from the later novel: most notably, this is no romance between doctor and patient but the story of a professional relationship between a woman doctor and her male nurse, an ex-convict who desires to "save lives" without compensation to expiate his sins.

What "Zerviah Hope" does accentuate about *Doctor Zay* is, first, the fundamental, curative dynamic between an idealized woman doctor and a man brought low by his own nature; second, in both narratives, the basic medical paradigm is one of nerves—the illness or lowness is a matter of character, but it expresses itself as nervousness and is treated homeopathically by the exercise of natural sensitivity, empathy, submission, and "feminine power" (81). Despite being in touch with his sensitive side, Zerviah is explicitly manly. We learn that his crime had been one of passion, defending the honor of a beautiful and wronged woman (83). And when he confesses his crime to Dr. Dare, we're told that he wants her to feel that he is "man enough to bear the burden of his deed" (83). Dr. Dare, meanwhile, is not all androgynous professionalism; her womanliness has not been leeched out of her by her work. She responds to Hope briskly and professionally, but we're told that there was "unprofessional moisture on her long, feminine lashes" (84). Later, when she is still recovering from a bout of fever, she seeks Hope "nervously" and breaks down in public sobs when he dies (88).

Though in typical nineteenth-century fashion, Zerviah the reformed sinner does die, his reformation and absolution occur at the hands of the woman doctor who initially puts faith in him when other townspeople won't, and finally blesses him—"her hand upon his eyes"—and allows him to rest in peace (88). Of Zerviah Hope the narrator comments: "A ruined character may be as callous as a paralyzed limb. A ruined and repentant one is in itself an independent system of sensitive and tortured nerves" (84). In "Zerviah Hope," then, Phelps uses neurological metaphors to talk about character reformation and forgiveness. As Zay does for Yorke, Dr. Dare helps to cure this heartsick man, though here the patient is more metaphorically than actually a system of hypersensitive and tortured nerves. Yorke's nervous condition in *Zay* is literal, though not without the same metaphorical implications (his character will be reformed). And, of course, there is no possibility in "Zerviah Hope" that the woman doctor will enter into a romance with an ex-convict; his death, therefore, makes for a neat resolution to his story.

In contrast, in *Doctor Zay*, romance eventually displaces herbal tonics in Zay's relationship with Yorke. As with the tonics, though, romance is administered homeopathically—in small and gently stimulating doses. She largely withholds herself, even sending her patient back to Boston twice, and so forces Yorke's own system to strengthen itself. In another narrative, one might read this as traditionally coy feminine behavior; in Phelps's novel, though, Zay's resistance of Yorke is motivated by honest clinical beliefs and sound professional self-defense. As Yorke becomes physically stronger and regains a sense of his manliness, he becomes aggressive in his courtship of her, and though she is not immune to his pursuit, she nevertheless believes his declarations of love to be "pathological sentiment" (211). Separation will be best for his continuing "spinal symptoms": "In a few months," she tells him, "you will find it easier to let me alone than to shatter your nervous system over me in this way. . . . [Y]ou will rebound to your own environment as naturally as you will walk without crutches. I have been nothing but a crutch to you, Mr. Yorke" (210).

That Yorke return to Boston is necessary for Zay as well. She will accept for herself nothing less than an "absolute marriage" of equals that allows her to continue in her profession on her own terms. Yorke, as she continually points out to him, is not "strong" (advanced) enough to love her in this way. As a physician, cognizant of Victorian theories of sexuality, Zay would have understood all too well the "old style" man's desire to conquer a woman; to accept Yorke's admiration at this stage in his development would, therefore, place her own heart and her career in jeopardy. Indeed, at this stage, he fluctuates between telling her that he would never expect her to give up her career and accusing her of having ruined "all that is noblest and sweetest and truest" in her womanly nature by becoming a doctor (210). Such a charge is very much in keeping with the criticisms launched by Weir Mitchell and other "old men" that women were naturally unfit to study and practice medicine and that they would damage their health and "unsex" themselves by doing so.[56]

With regained physical health, Yorke threatens to forget the female sympathy learned in the sickroom and to regress into the traditional old manliness romanticized by Weir Mitchell and conservatives in nineteenth-century culture. During his first return to Maine, the now physically healthy Yorke, "stronger, by far" than he had ever been in his life, attempts to conquer Zay by forcing her to admit that she loves him (227). Yorke has Zay at a disadvantage here for although he is a "well man," she has been weakened by diphtheria and overwork (227). Playing upon her culture's stereotypes of appropriate male and female sexual behavior, Phelps portrays their confrontation as that of a hunter and his prey. In a brown dress trimmed in "leopard fur," Zay appears "a beautiful, wild creature" (227). Yorke wonders, "Who was he that he should think to tame her? Yet, should a man let go his hold on a moment like this?" (227). Yorke does manage to wrestle a declaration of love from Zay and to force her to admit that she is overwhelmed by his newly acquired manliness—a manliness that is her "own work," Yorke declares (240).

Yorke's "manly" behavior here and his desire to conquer Zay romantically and sexually correspond to the ideal of healthy manhood that Weir Mitchell sought for his male patients. Though Mitchell believed that men needed to balance the effort of manly will against the danger of nervous exhaustion, he nevertheless argued that "constantly to struggle in a manly way is sure to bring nearer the day of perfect self-conquest."[57] Phelps, however, rejects Mitchell's ideal of manhood. Zay makes clear to Yorke that to be conquered, even in love, is not what she wants: "'You do not understand how to talk to a woman!' blazed Doctor Zay. 'It is presumptuous. It is unpardonable. You torture her. You are rough. You have no right'" (229). She sends him away again and, for months, will respond to his "deep" letters only with simple, straightforward "camaraderie" (249).

When he returns to Maine for the last time, Yorke has reached the higher plane of New Manliness. He comes not to conquer but to "rest" her, and he speaks to her in a way that "gave to his urgency more the force and calm of an assured, long-married love than of a crude young passion. . . ." (256). Having been strengthened in mind, body, and spirit through the offices of Doctor Zay and her homeopathy, Yorke is decidedly "manly," though he retains the rudiment of his feminine sympathy. No longer trying to force his manliness upon her, Yorke is complacent when she, for instance, criticizes his driving and takes the reins from his hands: "'I don't care who has the reins,' he cried with a boyish laugh, 'as long as I have the driver!'" (257). When Zay seems to bend toward agreeing to marry him, he does not press his advantage but asks her to come to him willingly: "'I don't want to feel as if I were taking a sort of – advantage. If you put me off one minute longer, I – shall.

I shall take all I can get. I shall like to remember, all my life, that you came to me first, of your own accord; that you loved me so much, you would grant me this—little proof'" (258). Zay, who has fallen in love with this creature of her own making, consents to his request with the understanding that she will remain a "strong-minded doctor," and the reader is left to imagine the success of their experiment in absolute marriage (254).

"You see, Mr. Yorke," Zay tells her patient, "you have been so unfortunate as to become interested in a new kind of woman. The trouble is that a happy marriage with such a woman demands a new type of man" (244). Yorke's ability and willingness to accommodate himself to marriage on Zay's terms is not just a statement about the sexual politics of absolute marriage, the marital hope for the New Woman. It also argues for the acceptance of a new image of the public sphere, one in which professional women work alongside professional men. Weir Mitchell, Edmund Clarke, and other mainstream physicians argued forcefully against such professional partnership when they asserted that women were "making mischief" with their health and, by extension, with the health of the race by their "desire to be on a level of competition with man and to assume his duties." Phelps, however, sees not neurological mischief but new hope for the race in the figure of the ideal woman and, especially, in the woman homeopath. What Mitchell and mainstream culture at the turn of the century saw as an undermining of manhood by the professional New Woman, Phelps shows to be a tonic to aid true and New Manliness.

Notes

1. See Elizabeth Segal, "The Gypsy Breynton Series: Setting the Pattern for American Tomboy Heroines," *Children's Literature Association Quarterly* 14 (Summer 1989), 67.
2. Qtd. in Mary Angela Bennett, *Elizabeth Stuart Phelps* (Philadelphia: University of Pennsylvania Press, 1939), 1.
3. Elizabeth Stuart Phelps, "To Silas Weir Mitchell," 25 January 1884, qtd. in Jennifer Tuttle, "Letters from Elizabeth Stuart Phelps (Ward) to S. Weir Mitchell, M.D., 1884–1897," *Legacy: Journal of American Women Writers* 17.1 (2000), 85; Elizabeth Stuart Phelps, "To Silas Weir Mitchell," 3 February 1884, qtd. in Tuttle, 86.
4. Tuttle, "Letters," 83–84.
5. Qtd. in Frederick Wegener, "'Few Things More Womanly or Noble': Elizabeth Stuart Phelps and the Advent of the Woman Doctor in America," *Legacy* 22.1 (2005), 1–17, 4. Phelps is responding to Clarke's *Sex in Education* (1873).
6. Tuttle, "Letters," 86.
7. E. Anthony Rotundo, *American Manhood: Transformations in Masculinity from the Revolution to the Modern Era* (New York: Basic Books, 1993), 269.

8. Barbara Will, "The Nervous Origins of the American Western," *American Literature* 70.2 (June 1998), 293.
9. See Anne Kirschmann, *A Vital Force: Women in American Homeopathy* (New Brunswick: Rutgers University Press, 2004).
10. Martin Kaufmann, *Homeopathy in America; the Rise and Fall of a Medical Heresy* (Baltimore: Johns Hopkins Press, 1971), 25.
11. John S. Haller, *American Medicine in Transition, 1840–1910* (Urbana: University of Illinois Press, 1981), 108. See James Tyler Kent, *Lectures on Homeopathic Philosophy* (1900; Richmond, California: North Atlantic Books, 1979) for homeopathic tenets that are roughly contemporary with *Doctor Zay*.
12. Qtd. in Tuttle, "Letters," 86.
13. See Anne Stiles's introduction to this volume.
14. Jane Wood, *Passion and Pathology in Victorian Fiction* (New York: Oxford University Press, 2001), 28.
15. See, for example, his short story, "Autobiography of a Quack," which first appeared in the *Atlantic Monthly* in October and November of 1867.
16. *"Halfway Up the Hill," Women Healers and Physicians: Climbing a Long Hill*, ed. Lilian R. Furst (Lexington: University Press of Kentucky, 1997), 234.
17. Regina Morantz-Sanchez, *Sympathy and Science. Women Physicians in American Medicine* (New York: Oxford University Press, 1985), 31.
18. Sarah Grand, "The New Aspect of the Woman Question," *North American Review* 158 (1894), 270–7, 271.
19. Grand, "New Aspect," 271. According to Carolyn Christensen Nelson, the phrase New Woman started being capitalized within two months of its being picked up in the press. "Introduction," *A New Woman Reader: Fiction, Articles, and Drama of the 1890s* (Peterborough, Ontario: Broadview Press, 2001), ix.
20. Olive Schreiner, *Woman and Labour* (1911; London: Virago Press, 1978).
21. Grand, "New Aspect," 272. See also Grand's essay, "The Man of the Moment," published a few months later in the *North American Review* 158 (1894), 620–8.
22. Wood, *Passion and Pathology*, 63–4.
23. Morantz-Sanchez, *Sympathy and Science*, 232.
24. Morantz-Sanchez, *Sympathy and Science*, 6.
25. See Kristine Swenson, *Medical Women and Victorian Fiction* (Columbia: University of Missouri Press, 2005); Lilian R. Furst, ed. and introd., *Women Healers and Physicians: Climbing a Long Hill* (Lexington: University Press of Kentucky, 1997).
26. William Dean Howells, *Dr. Breen's Practice, Atlantic Monthly* 48 (1881), 592.
27. Michael Sartisky, Afterword, *Doctor Zay* by Elizabeth Stuart Phelps (1882; New York: The Feminist Press, 1987), 302.
28. W.D. Howells to Elizabeth Stuart Phelps, 28 October 1881, in the *Atlantic Monthly* 49 (1882), 294, 518.
29. See Sartisky, Afterword, *Doctor Zay*, 300.
30. Wegener, "'Few Things More Womanly or Noble,'" 16.
31. Elizabeth Stuart Phelps to Silas Weir Mitchell, 18 November 1884, qtd. in Tuttle, "Letters," 87–8.
32. See, for example, Sarah Orne Jewett's *A Country Doctor*, Henry James' *The Bostonians*, or Charles Reade's *A Woman-Hater*.
33. Qtd. in Wegener, "'Few Things More Womanly or Noble,'" 4.

34. "Packaging the New Woman," in *The New Woman in Fiction and in Fact*, eds. Angelique Richardson and Chris Willis (New York: Palgrave Macmillan, 2001), 56.
35. Tuttle, "Letters," 86.
36. Elizabeth Stuart Phelps, *Doctor Zay*, 161. Further references to this source will be parenthetical.
37. According to the *Oxford English Dictionary*: "Path. Excessive and morbid sensitiveness of the nerves or nerve-centres. 1880 M. MACKENZIE Dis. Throat & Nose I. 415 Hysterical persons, suffering from hyperaesthesia or paraesthesia of the larynx, often erroneously fancy that something is sticking in the part." Hyperaesthesia was considered related to both neurasthenia and hysteria. David Payne claims it to be an "obvious precursor of Freudian conversion hysteria" in *The Reenchantment of Nineteenth-Century Fiction. Dickens, Thackeray, George Eliot and Serialization* (New York: Palgrave Macmillan, 2005), 113. Gillian Beer calls hyperaesthesia "the particular disease of Victorian consciousness" (qtd. in Payne, 113).
38. Silas Weir Mitchell, *Wear and Tear or, Hints for the Overworked* (1871), http://www.gutenberg.org/files/13197/13197-h/13197-h.htm (accessed 21 January 2007).
39. See Tom Lutz, *American Nervousness, 1903: An Anecdotal History* (Ithaca: Cornell University Press, 1991); E. Anthony Rotundo, *American Manhood: Transformations in Masculinity from the Revolution to the Modern Era* (New York: Basic Books, 1993); Angus McLaren, *The Trials of Masculinity: Policing Sexual Boundaries, 1870–1930* (Chicago: University of Chicago Press, 1997).
40. Will, "The Nervous Origins," 297.
41. McLaren, *Trials of Masculinity*, 155.
42. McLaren, *Trials of Masculinity*, 155.
43. See Angelique Richardson, *Love and Eugenics in the Late Nineteenth Century. Rational Reproduction and the New Woman* (New York: Oxford University Press, 2003).
44. Will, "The Nervous Origins," 294.
45. According to Kevin White, who examined actual marriages between New Women and New Men, Zay has good reason to worry about the viability of a marriage to Yorke. Of the nine cases that White details, four of the marriages end badly, with the wife suffering most from the experience. White places the blame for the failed experiments on "the problem of the New Man" (*The First Sexual Revolution: The Emergence of Male Heterosexuality in Modern America* [New York: New York University Press, 1993], 107).
46. Aconite pellets are derived from monkshood and remain standard homeopathic treatment for a variety of symptoms including "sudden illness, aches, colds, coughs, sore throat, flu, chills and fever" though "fear and restlessness seem to dominate the symptoms" (Elixirs.com, Homeopathy for Health [2005], http://www.elixirs.com/products.cfm?productcode=S61A [accessed 21 January 2007]).
47. Elixirs.com, Homeopathy for Health (2005), http://www.elixirs.com/products.cfm?productcode=S61A (accessed 21 January 2007).
48. Thomas Keller, "Railway Spine Revisited: Traumatic Neurosis or Neurotrama?," *Journal of the History of Medicine and Allied Sciences* 50 (October 1995), 507–24, 511.

49. See Tuttle, "Letters," 89.
50. Zay's use of brandy, in particular, shows her to be an "eclectic" homeopathic practitioner rather than a strict follower of Hahnemann's original homeopathic principles. In her adoption of certain nonhomeopathic treatments that were "time-tested" by allopaths, Zay is typical of American homeopaths of her era rather than unusual (Kaufman, *Homeopathy in America*, 113). Kaufman and others have noted as well that both homeopaths and allopaths were borrowing treatments and methods from one other and were, to some extent, meeting in the middle of their earlier therapeutics battle. Nevertheless, many mainstream physicians, including Weir Mitchell, remained contemptuous of homeopathy even as they adopted certain homeopathic treatments.
51. "Nux vomica," Homeopathic Library (23 November 2006), http://www.internethealthlibrary.com/hom-library/nux-vomica.htm (accessed 21 January 2007).
52. Charles Darwin, *The Descent of Man* (1871) in *Darwin*, ed. Philip Appleman (New York: Norton, 1979), 132–208, 140.
53. Geraldine Jewsbury to Jane Welsh Carlyle, Letter 96 (undated–1849?), from *Selections from the Letters of Geraldine Endsor Jewsbury to Jane Welsh Carlyle*, ed. Mrs. Alexander Ireland (London: Longmans, Green and Co., 1892).
54. Silas Weir Mitchell, *Doctor and Patient* (1901), http://www.truthbeknown.com/doctor_and_patient.html (accessed 21 January 2007).
55. Elizabeth Stuart Phelps, "Zerviah Hope," *Scribner's Monthly* 21.1 (November 1880), 78–89.
56. See Swenson, *Medical Women*, 40.
57. Silas Weir Mitchell, "Clinical Lecture on Nervousness in the Male," *The Medical News and Library* 35 (December 1877), 178.

III. Sex and the Brain

5

Trauma and Sexual Inversion, circa 1885: Oliver Wendell Holmes's *A Mortal Antipathy* and Maladies of Representation

Randall Knoper

A mysterious young man appears in a late-nineteenth-century New England village and inexplicably avoids the young women there, preferring the sole company of his Italian manservant and therefore arousing great curiosity. His secret, as it eventually comes out, is rooted in a childhood shock with sexual overtones and powerful sexual effects. As a baby, while being held and tossed by his beautiful and nubile cousin, he was dropped off a balcony into a thorn bush. The lasting effect of this trauma is that, whenever he is in the presence of a young woman exuding a sexual aura, the natural outward impulse of his nervous system is inverted, turns inward, and arrests his heart. Heterosexual arousal is fatal, and Maurice Kirkwood is thus understandably gynophobic, with an aversion for young women powered by an antipathic neurological response.

This is the plight of the main character in *A Mortal Antipathy*, published by Oliver Wendell Holmes in 1885.[1] Although the novel has fallen into the forgotten territory of decanonized literature, its early, prescient linking of trauma and a kind of sexual inversion demands our attention for what it can tell us about the pathologizing of homosexuality; it inaugurates a strain of thought that still plagues us. I should say at the outset that this fictional case is not one of same-sex desire. But it *is*, quite literally, a case of inverted sexual impulse, of "sexual inversion"—the term used to describe homosexuality. Given that "antipathic sexual instinct" was also used to describe homosexuality, I will argue that Kirkwood's mortal antipathy belongs in the amorphous territory of sexual deviance that included homosexuality and that the sexology of the 1880s sought to explain. This fictional case helps restore to our historical sense of sexual inversion—and our understanding of the pathologization of homosexuality—a meaning of literal neurological inversion, or reversed nerve force; the notion of an "antipathic instinct," as a repulsive neural force akin to the opposite of magnetic attraction, further fleshes out this understanding. But while Kirkwood's fictional case helps clarify the historical conceptions of sexual inversion as a neurological disease, its causal etiology more importantly casts light on the earliest linkage between trauma and supposed sexual deviance. Our pursuit of this link will lead down unexpected routes, into matters of representation and reproduction and into the intertwinings of art and the rapidly developing neuroscience of the late-nineteenth century—two areas that Holmes, a poet as well as a Harvard professor of anatomy and physiology, persistently joined.

The novel appeared at a moment when theories of trauma—including the mysteries of psychological trauma—and theories of sexual inversion were proliferating, becoming central concerns of neurologists and alienists, but curiously developing independently of each other. Holmes, seemingly for the first time, brought the two together. No one else appears to have been thinking along the same lines. It was not until 1905 that Sigmund Freud published his *Three Essays on the Theory of Sexuality*, perhaps the most influential diagnosis of "contrary sexual feelings" and "inversion" as the result of a distressing sexual experience in early childhood.[2] If our history of the pathologizing of homosexuality as an effect of childhood psychological trauma takes Freud as its strongest point of origin, Holmes's earlier linkage nonetheless tells us something important and different. Specifically, the inspiration behind this combination lay, I will argue, in the connection Holmes made between trauma and sexual inversion as disorders of representation and reproduction, which he thought of analogously. Holmes's innovation provides an instance, that is, of the subtle intertwinings of science and the concerns and metaphorical thinking of art, the ways one can provide grounds for conceiving the other, and the ways the two can join to construct a cultural latticework. Here, thinking metaphorically about representation and reproduction brings about a new etiological explanation at the same time that it helps define a pathology and fix attitudes about sexuality. In turn, neuroscience provides scientific particulars that pathologize sexual inversion as a malady of expression. *A Mortal Antipathy*, then, serves as a focus and an occasion for unraveling a knot that binds together discourses of shock and sexual deviance, the science of the nervous system, and questions of artistic and biological expression—all tight-knit in an ideological configuration that cast its shadow far into the future.

Shock and memory circuits

Before reaching for these conjunctions, though, an overview of the separately developing understandings of psychological trauma and sexual inversion will help situate Holmes's story. And the history of late-nineteenth-century conceptions of psychological trauma must come first. The most familiar reference point for situating the malady of Holmes's protagonist would probably be Sigmund Freud and Josef Breuer's work of the early 1890s, which culminated in *Studies in Hysteria* in 1895—an apt framework for thinking about *A Mortal Antipathy* because it made a causal link between childhood sexual trauma and neurosis, specifically hysteria. But of course this work was published some years after Holmes's fictional case, so we must look earlier and elsewhere for context. And in the early 1880s, notions of psychological trauma were emerging especially from the controversy that swirled around "railway spine"—an ailment for which train-accident victims could be granted thousands in damages—and around the questions that railway spine foregrounded about physical and emotional trauma, their differences and their similarities.[3]

The British surgeon John Eric Erichsen, in 1866, had defined railway spine as a physical injury, a supposed "concussion" of the spine resulting in an indiscernible "molecular disturbance"—without a visible lesion—that issued in a range of posttraumatic maladies. His characterization of railway spine was used to support lawsuits against the railroads because it connected the mysterious symptoms many plaintiffs experienced to a concrete bodily injury supposedly suffered in railway accidents.[4] Controversial from the start, this idea of spinal concussion was notoriously and influentially discounted in the early 1880s by British railway surgeon Herbert Page, first in an essay that won Harvard University's Boylston Prize and then in his expanded study, *Injuries of the Spine and Spinal Cord Without Apparent Lesion, and Nervous Shock, in their Surgical and Medico-Legal Aspects* (1883). Page's arguments against the existence of spinal concussion—including his suggestion that the disorders of railway spine may have begun in a trauma-induced hypnotic state that caused dazed victims unconsciously to mimic real physical illnesses, as hysterics did—were used to undermine injury lawsuits. While his study discredited the idea that physical concussion caused railway spine, however, Page brought to the fore the ideas that psychological shock, instead of physical spinal injury, might be the cause of traumatic malady; that "great fear and alarm" alone might cause the common symptoms of shock and might even cause death; and that, indeed, "purely mental causes" might be the best explanation for symptoms that appear in the absence of any apparent physical injury, especially for symptoms that surface some time after an accident.[5] Page certainly was not the first to make the argument that psychological shock can cause physical maladies, but his work became the touchstone for subsequent investigation of the effects of emotional trauma and a locus for thinking about the relation between mind and body.[6]

This emerging argument about trauma, and the possible physical effects of emotional shock, was joined in 1883 and 1884 by Holmes's Harvard colleague, the Massachusetts General Hospital neurologist James Jackson Putnam, along with another physician at the hospital, George L. Walton. They agreed with Page that mental or emotional states could cause posttraumatic symptoms, but they also forwarded the hypothesis that these symptoms were hysterical and, since railway spine most often affected men, that these were cases of male hysteria—or neurasthenia, to avoid the derogatory womanly connotations of hysteria.[7] Putnam and Walton were strongly influenced by Jean-Martin Charcot, whose theories of hysteria were certainly the most prominent at this time, and Charcot in turn cited their work in his own research, as he turned from studying only female hysterics to investigating male hysteria, and from defining hysteria solely in terms of heredity to considering the possibility—tentatively, at least—that it might be psychologically and traumatically triggered. Charcot made his first diagnoses of male hysteria in 1879, published his first article on the topic in 1882, and by 1885, at least, was grappling with cases he believed were induced, if not caused, by emotional trauma.[8] If he affirmed the existence of male cases precipitated by terror or fright, however, he nonetheless tried to adhere to his somatic model, suggesting that such trauma is still shock in a very literal sense—that is, it translates as electric shock, an electrical impulse that has a physical effect on the nerves.[9] If spinal concussion is not credited here as the physical cause of posttraumatic symptoms, emotional shock itself becomes a physical cause. Somewhat differently, however, in an argument seemingly spurred by Page's theorizing about trauma victims' hypnotically unconscious mimicry of others' injuries, Charcot in 1884 and 1885 argued that traumatic paralysis differs from organic paralysis because it is caused by an idea, or chain of ideas, set in motion outside consciousness, through hypnosis or suggestion.[10] Such a notion had its roots in British psychophysiology and its concept of an "ideo-motor" reflex arc, which occurs as a kind of physical brain–body activity outside consciousness. Designed to account for unconscious actions and symptoms in physiological terms, this kind of concept nonetheless straddled the divide between physical and psychological explanations for traumatic disorders. Not due to an organic injury, produced by an unconscious idea, this kind of trauma symptom still operated physiologically.

Making his fiction part of this medical dispute about the relation between emotional trauma and bodily ailment, Holmes devotes substantial space in his novel—through his mouthpiece, the wise Dr. Butts—to discussing psychological shock as the sole cause of physical maladies. Considering the question of physical versus psychological etiologies of traumatic disturbances, Butts, in accord with Page, declares that a "sudden mental shock" may cause insanity; our asylums can supply many such cases. Even worse, people may die of terror, may be "*scared to death*, literally." Emotional events do translate into physical states, and, Butts says, "as a violent emotion caused by a sudden shock can kill or craze a human being, there is no perversion of the faculties, no prejudice, no change of taste or temper, no eccentricity, no antipathy, which such a cause may not rationally account for." Musing on Kirkwood's antipathy before he learns the full truth, Butts thinks to himself that "some early alarm" may have "broken some spring in this young man's nature, or so changed its mode of action as to account for the exceptional remoteness of his way of life" (*MA*, 91–2). In line with Page, Putnam, and Walton, then, Holmes endorses the idea that shock in the sense of emotional or psychological trauma—a fright, an alarm—can be the cause of multiple pathologies and can have material bodily effects, from creating an antipathy to causing death.

But in what we might consider an elaboration of Charcot's distinction between traumatic paralysis and organic paralysis, Holmes then makes a distinction in the causal mechanism between simply physical and psychological origins. Talking about antipathies and aversions, Butts says that when people faint at the sight of blood, or have spasms or fits of sweating when they hear certain sounds, such cases of "profound disturbance of the nervous system" are "produced by impressions on the organs of sense, seemingly by direct agency on certain nerve centres" (*MA*, 89). It is as if sensory perception is a kind of physical agency whose force makes a material impression directly on a nerve center, causing the "disturbance" as if it were a sensorimotor reflex action—in a way similar to Charcot's materialization of shock as an electrical impulse. Butts distinguishes, however, between this and "another series of cases in which the imagination plays a larger part in the phenomena" (*MA*, 89). Peter the Great, as an infant, fell from a bridge into the water, and ever after dreaded the sound of wheels on a bridge. Another man, when an infant, had a shock when the cord supporting a weight in a clock broke, and the weight came crashing down in the case; ever after, when he passed a tall clock, he experienced profound nervous agitation. While Butts still speaks of the initial trauma in these cases in terms of material mechanism, as instances in which "Some effect must have been produced upon the pulpy nerve centres from which they never recovered," the differentiating function for these victims, as for Maurice Kirkwood, is that of imaginative association—between a remembered trauma and wheels on a bridge or clocks or young women (*MA*, 91). A present perception triggers a memory associated with the trauma, and the original nervous turbulence is reproduced. Notably, Holmes's version of this associationism is physiological. Articulated in the vocabulary that had developed in the 1860s of neural pathways, the conception of imaginative association bears importantly on the prospects for curing Kirkwood.

But the complications in Holmes's fictional effort to assay the kinds and effects of shock get additional turns of the screw that go beyond this uncertain territory of mind–body intermingling. Specifically, he invokes two theories about the effect of shock on bodily energy—its stifling of vital force or its reversal of nervous energy—that more particularly describe Kirkwood's sexual economy. Both theories draw upon understandings of *inhibition* as part of the regulatory operation of the nervous system. In the first, shock causes a reflex depression of vital force. Ever since the eighteenth century, as historian Roger Smith notes, there had been a notion of "spinal shock"—in which shock works to inhibit reflex action, so that in the case of a decapitated frog, the animal's spinal system would not resume reflex function until some time after the brain had been destroyed.[11] Herbert Page had similarly entertained the idea of shock as a "reflex paralysis" or "reflex inhibition" that involved "a lowering of the vitality of every organ and function of the body" (*ISSC*, 144–5). Such reflex inhibition suits the condition of Maurice Kirkwood, not only because the reflex action of his heartbeat is stayed by the shock, but also because he suffers the "sudden collapse of all the powers of life" (*MA*, 231). In these terms, shock might be an inhibition simply in the sense of a reflex depression of nerve force, a lowering of vital force, a stanching of the well of life.

But another conception of shock as causing a *reversal* of nerve force surfaces in these discussions of the 1880s. British surgeon John Furneaux Jordan, invoking the theory of the correlation and convertibility of force, argued that shock could not be a depression of nerve force, because "no force can be depressed; it can only be metamorphosed into some other of the numerous forces which prevail, or are capable of prevailing, in the animal economy." Shock, then, "is essentially a metamorphosis of nerve-force" (*SASO*, 42).[12] Page endorsed Jordan's idea and further suggested that, in a metamorphic *reversal* of force, people with greater nerve force will suffer greater effects of shock (*ISSC*, 154–6); the abundance of nervous energy in such victims is not simply stifled but is actively turned against them as an inhibitory force. In an elaboration of the idea, which resonates especially well with Kirkwood's case, Jordan focuses on the effects of shock on the heart, which he considers a primary focus and gauge of shock (*SASO*, 22). In his conception—attuned to the advances being made in the 1880s, especially by physiologists at Cambridge University, in understanding the regulatory system of the heart—nervous reflexes both excite and inhibit the heartbeat, acting in an equilibrium of antagonistic forces to control heart rate.[13] Specifically, the "pneumogastric" nerves "retard or inhibit cardiac action," while "the sympathetic stimulates the heart to increased action" (*SASO*, 44). Just as nerve force, then, can be metamorphosed from an inspiriting agency to a depressive one, so can it both stimulate heart rate through the sympathetic nerves and slow or arrest the heart through the pneumogastric nerves.

These conceptions of shock suit the condition of Maurice Kirkwood, partly because it is his reproductive impulse, the fundamental "spring" of vitality, that is depressed, and partly because that strong impulse has indeed been metamorphosed through shock into an equally strong reversed force, an imbalanced "reflex inhibition" that has "so changed its mode of action" as to turn its energy against life, to snuff it out rather than help it bloom. Even more particularly, a medical report on Kirkwood to the Royal Academy of the Biological Sciences asserts that his case is one of "the effect of inhibition on the heart." In a kind of summary of the then-current knowledge about reflex regulation of the heartbeat through excitation and inhibition, the report explains that one group of nerve cells "acts as the spur" on the heart, accelerating it, and another "as the bridle," retarding it. A "disturbed" influence of "the centre of inhibition"—as "the restraining influence over the heart is called"—is the usual cause of "well-known cases of deadly antipathy" (*MA*, 235–6). In accord with these understandings of shock, Kirkwood's initial emotional trauma exaggerates an inhibitory reflex action that arrests his heart. A nerve force that normally operates in a regulatory balance with the enlivening stimulus gains the upper hand. And now the presence of any young woman opens the associational pathway that reactivates this inhibitory, death-dealing power opposing vitality itself.

The final significant context for understanding the operation of Kirkwood's traumatic memory is the later nineteenth-century reconceptualization of association. Associationism, of course, had been one of the dominant philosophical-psychological conceptions of how the mind worked, stemming especially from the writings of John Locke and then James Mill. In the latter half of the nineteenth century, various theorists, notably Alexander Bain and Herbert Spencer, worked to "physiologize" associationism, attempting to match conceptions of mental association to developing knowledge about the nervous system and the brain.[14] As neural pathways were traced in the 1860s and 1870s, this anatomical knowledge combined with understandings of electrical science that allowed modeling of mental associations and memory in terms of intense or repeated electrical flows down neural pathways. Théodule Ribot's *The Diseases of Memory* (1881; English trans. 1883), synthesized and summarized this physiological thinking about memory and became a touchstone. For Ribot, memory was "essentially a biological fact, and only by accident a fact of psychology," and more particularly, it was "not a collection of impressions, but an assemblage of dynamic associations."[15] These associations, which were physical flows between nerve cells, he compared to muscular actions, which become ingrained habits over time; repetition likewise established the persistence of memory, as the flows of association established their traces, thereby encouraging nerve force to stream down their channels. As Francis G. Gosling observes, "Images of ruts, channels, and paths appear regularly" in the neuroscience of this period, "based on the concept that repeated actions along certain nerve paths eventually made it unlikely that impulses would travel any other route, much like the process of feet wearing a path across a lawn." Such a worn path constituted a durable memory. Assuming, relatedly, that pathologies were due to "morbid" channelings or "pernicious" pathways, therapies were developed to disrupt such habitual neural functioning—by, say, a change of scenery, which could establish new associations and pathways.[16]

Maurice Kirkwood's fatal association between his childhood trauma and sexually alluring young women conforms to these physiological ideas about memory. After the accident, when his cousin Laura approached him, Kirkwood reports, "the dread that she was about to lay her hand upon me had called up the same train of effects which the moment of terror and pain had already occasioned"—which included the arrest of his heart (*MA*, 209). As he further explains, "It was too evident that a chain of nervous disturbances had been set up in my system which repeated itself whenever the original impression gave the first impulse"

(*MA*, 211). And the original impression extends beyond Laura to all women of her age. Because Kirkwood will feel mortal effects when he is simply near a young woman, even if he does not know she is there, Dr. Butts hypothesizes that she may emit an electrical or magnetic "effluence" that triggers the mortal response (*MA*, 242). Holmes, however, in his preface to the novel, explains that there are especially strong associations between the olfactory nerves and those of "the higher organs of consciousness." In Kirkwood's case, the odor, or "atmosphere," of his sexually ripe cousin—and of any woman of her age and condition—became associated with his traumatic fall. As Holmes puts it: "It was an atmospheric impression of this nature which associated itself with a terrible shock experienced by the infant. . . . The impression could not be outgrown, but it might possibly be broken up by some sudden change in the nervous system effected by a cause as potent as the one which had produced the disordered condition" (*MA*, 217). Kirkwood is left "with an unconquerable fear of woman at the period when she is most attractive not only to adolescents, but to children of tender age, who feel the fascination of her flowing locks, her bright eyes, her blooming cheeks, and that mysterious magnetism of sex which draws all life into its warm and potently vitalized atmosphere" (*MA*, 213); as the report to the Royal Academy of Biological Sciences notes, he has no such reaction to "an old withered crone" and has a "timid liking for little maidens" (*MA*, 231). Finally, that report elaborates on his case in a way that echoes conceptions of associations as neural pathways, and that explains how this neural connection between trauma and young women then works to stop Kirkwood's heart:

> A single impression, in a very early period of atmospheric existence, . . . may establish a communication between this centre [the center of inhibition in the brain] and the heart which will remain open ever afterwards. How does a footpath across a field establish itself? Its curves are arbitrary, and what we call accidental, but one after another follows it as if he were guided by a chart on which it was laid down. So it is with this dangerous transit between the centre of inhibition and the great organ of life. If once the path is opened by the track of some profound impression, that same impression, if repeated, for a similar one, is likely to find the old footmarks and follow them. Habit only makes the path easier to traverse, and thus the unreasoning terror of a child, of an infant, may perpetuate itself in a timidity which shames the manhood of its subject. (*MA*, 236)

The report leaves open the possibility that there can be a remedy for this perverted current, a redirection of its flow into "natural" channels—which, as we shall see, leads to Kirkwood's cure. But first let us attend to the other context for this condition, the heterosexual timidity that shames one's manhood, the sexual dysfunction associated with Kirkwood's odd behavior.

Inverted sexual feeling

None of our histories of the invention and psychologization of trauma has anything to say about sexual inversion—theories of which, I will argue, inform Holmes's representation of Kirkwood's gynophobia. Sexual inversion, rather, fits into another history, now also beginning to seem solidified and familiar, a history of the medicalization of homosexuality.[17] As such scholars as Harry Oosterhuis and Vernon A. Rosario tell it, the important moments in this history include the 1869 article on *conträre Sexualempfindung* by Karl Westphal, the editor of the German *Archives of Psychiatry and Nervous Diseases*, which was followed by a flurry of publications in Germany on "contrary sexual feeling," culminating in an 1877 essay on the subject by Richard von Krafft-Ebing—who would call his compendious life-work *Psychopathia Sexualis, with Especial Reference to the Antipathic Sexual Instinct* (1886).[18] A group of physicians in the United States also wrote on the topic in the early 1880s.[19] In a way similar to studies of hysteria written in the late-nineteenth century, these essays on "perverted sexual instinct" tend to refer to their topic as a hereditary neuropsychopathic condition, not as something acquired. And they mix various kinds of cases together—those of clear same-sex erotic attraction, but also instances of gender-identity inversion and transvestism. Exactly what antipathic or inverted sexual instinct meant is often unclear, as are the specific etiologies. An "organic brain defect," "congenital perversion of the sexual instinct," "a disease of the central nervous system," "hereditary taint," a "perversion somewhat analogous to hysteria," a nervous father, a hysterical mother— all these conditions or causes are mentioned as ways of explaining the patients' sexual deviations.[20]

In 1882, Charcot and his colleague Valentin Magnan entered the discourse with their essay "Inversion du sense génital" (building on the conception of "inversion of the sexual instinct," introduced in 1878 by the Italian doctor Arrigo Tamassia).[21] Curiously, however, Charcot does not mingle his interest at that time in trauma with this new subject of sexual inversion.[22] The article apparently was written primarily by

Magnan, and its etiology of the "inversion of the genital sense" is invariably congenital and based on Magnan's theory of hereditary degeneration. Charcot never wrote again on inversion. Magnan did, developing an etiology of sexual disorders that located their causes at particular spots in the nervous system; the problems emanated from "primitive" cerebrospinal levels, thus making deviant sexual desires into degenerate, involuntary reflexes for which the sufferers were not responsible. This work instigated a flurry of interest by 1885, especially in the French medical journals, as systems were proposed to define and classify "antipathic" or "contrary" or "inverse" "sexual instinct" or "genital sense."[23] For the most part, the etiology remained that of hereditary, sometimes degenerative, taint. In a moment of near contact between theories of trauma and sexual inversion, while Magnan worked out his nosology of sexual deviance, he did entertain the possibility that mental disease generally could be due to trauma rather than inherited nervous system weakness.[24] But again, as with Charcot, the interest in sexual inversion and the interest in trauma did not make explicit connection.

One might argue that Holmes actually does not make the connection either, because the sexual inversion that Kirkwood's trauma leads to does not really suit the medical definitions, especially to the extent that they describe same-sex desire or explicit homosexuality. Kirkwood, after all, is *attracted* to young women. He says they "at once attracted and agitated me," but when near one there comes a sense of impending death, his circulation feels paralyzed, he feels faint (*MA*, 214–7). He *laments* his condition: "After all, what had I to live for if the great primal instinct which strives to make whole the half life of lonely manhood is defeated, suppressed, crushed out of existence?" (*MA*, 223). Also, Dr. Butts likens the problem to that of a young man who loses all confidence in the presence of the very woman to whom he is most attracted, and the report to the Royal Academy of Biological Sciences similarly wonders if, instead of being anomalous, Kirkwood's condition is "only the last term of a series of cases which in their less formidable aspect are well known to us in literature, in the records of science, and even in our common experience"—thereby seemingly drawing Kirkwood into the continuum, or bell curve, of the heterosexually normal (*MA*, 246, 231).

On the other hand, his malady "was a reversed action of the nervous centres,—the opposite of that which flushes the young lover's cheek and hurries his bounding impulses as he comes into the presence of the object of his passion" (*MA*, 215). And we are told that "the natural current of the instincts has had its course changed as that of a stream is changed by a convulsion of nature, so that the impression which is new life to you is death to him" (*MA*, 225). By calling the condition the *opposite* of the heterosexual norm, and by using the language of inverted "instinct" and antipathic instinct, Holmes seems to bring Kirkwood's case into the realm of sexual inversion, where the opposite of heterosexuality may be a death-drive reversal of the reproductive impulse, or an inverted "polarity" which, as the opposite of the male–female, plus-and-minus electrical attraction, suggests something other than heterosexual eccentricity. It is worth noting, too, that the medical articles on sexual inversion refer to men who had "a feeling toward females . . . of indifference or repugnance,"[25] a "*horror feminae*,"[26] a "disgust for women,"[27] "a real dislike for girls," "a horror of women,"[28] and so on. These men, unlike Kirkwood, also experienced same-sex desire. But their similar antipathy toward women draws Kirkwood into their orbit. One case, reported in 1883 by William A. Hammond, former surgeon general of the U.S. Army and president of the American Neurological Association, sounds very much like Kirkwood's type: the patient was "a young man of twenty years of age, a graduate of college, of an aesthetic turn of mind . . . who condemned himself to solitude, who fled from the world and exhibited the most decided repulsion, not only toward women in general, but towards all that could exhibit the least evidence of a feminine origin."[29] This young man also was excited by pictures of naked men. But again, all outward, public signs match him nicely to Kirkwood.

Moreover, for Kirkwood there is significant guilt and shame associated with his condition. When as an adolescent Kirkwood goes to Italy, to consider enrolling in a Catholic boys school in order to avoid women, he gladly accepts from a friend a medal, blessed by the Pope, which is supposed to draw out "original sin" as well as "evil and morbid tendencies," though finally it does not change his "unnatural condition" (*MA*, 218). What exactly might he be viewing in himself as an evil and morbid tendency, in need of religious repair? A bit more explicitly, Kirkwood is finally brought to reveal his secret "for the light it throws on certain peculiarities of human character often wrongly interpreted as due to moral perversion, when they are in reality the results of misdirected or reversed actions in some of the closely connected nervous centres" (*MA*, 223). Is this meant to say that his condition is not a moral perversion, such as same-sex desire might be thought to be, but rather a nervous malady? Or is it meant to say that whatever his condition, again perhaps something that looks like same-sex desire, it is not a moral shortcoming but rather a physiological malady? Although Holmes clearly wants to absolve his protagonist of responsibility for any "evil" or "moral perversion," he also seems to want to invoke, and account for, sexual differences that can be seen as such.

Finally, there is an echoing subplot about a female student who qualifies as a neuropathic gender invert, and who therefore throws Kirkwood's similar status into sharper relief. Lurida Vincent is a suffragist who declares that she doesn't like men much, because they are tyrants, and that women could get along better without them (*MA*, 94, 283). She is an advocate of "virile womanhood," of the Amazonian kind, though she herself is scholarly rather than virilely muscular (*MA*, 40). Indeed, the regular "over-excitement" of her brain tends to sap her "vitality" (*MA*, 77). She keeps her hair short, partly because "Her head used to get very hot when she studied hard," as if all the required nervous force were radiating into the atmosphere (*MA*, 124). Her penetrating curiosity about Kirkwood leads her to want to study medicine, and most particularly she wants "to study up the nervous system, and learn all about it," though there is the suggestion that she wants this knowledge to explain her own debilitated system, as well as to penetrate Kirkwood's (*MA*, 104, 126). We learn that Lurida's "brain had run away with a large share of the blood which ought to have gone to the nourishment of her general system," so she doesn't have the "womanly developments" that the heroine of the novel, Euthymia, has. In Lurida's case, "An impoverished organization carries with it certain neutral qualities which make its subject appear, in the presence of complete manhood and womanhood, like a deaf-mute among speaking persons" (*MA*, 255–6). In other words, Lurida's vital force is wrongly channeled into intellectual pursuits, thereby foiling its natural physical expression in her body; the simile of the deaf-mute is telling, because it figures Lurida as a failure in self-expression. She does have a voice, though, which is "penetrative, aggressive," and expresses "the corresponding traits of mental and moral character" (*MA*, 145). But if she succeeds in that kind of masculine self-expression, we also must note that her brain demands so much blood that she cannot even blush when embarrassed—and thus is deprived of the exemplary outward expression of unmediated and sincere female modesty (*MA*, 132). She is implicitly contrasted to Maurice: for example, she has a "bold, almost masculine" style of handwriting, while Kirkwood's is "not decisive" (*MA*, 118, 95). In voice, body, failure to blush, and habitual gesture—the natural physical expressions of the inner person—Lurida shows herself a failure in "complete womanhood." As a foil to Maurice Kirkwood, she mirrors his condition of deviantly channeled nervous energy, energy that does not move sexually outward toward new life. She also quite clearly represents a gender inversion. Paired as they are, Lurida and Maurice underscore this story as one of sexual inversion, even if neither is explicitly homosexual.

Maladies of representation and their cures

This, then, is the framework for *A Mortal Antipathy.* But if this context of scientific discourse about psychological trauma and sexual inversion helps us to understand Holmes's novel, it still does not explain why Holmes was "ahead" of cutting-edge neurology and psychology in thinking about, and joining, these matters—in pathologizing sexual inversion in a new way, as an acquired trait caused by trauma. Now, it may be apparent from the foregoing discussion that analogical connections between trauma and sexual inversion were ready at hand. If shock was conceived as a depression of nerve force or life force, or if it was conceived as an exaggeration of inhibitory nerve force over stimulative, then it could logically enough join sexual inversion as an upending or derailment of vitality, a foiling of the life force that would otherwise attain its ultimate goal in reproduction. The neural inhibition of shock and the reversal of the heterosexual impulse in inversion would be matched as antilife forces, a conclusion that is easy and familiar enough. But I think there are additional complications in Holmes's thinking about the matter which come from his bundling together notions of biological reproduction, the nature of life to express itself, and the natural truthfulness of such expression when it follows its "proper" channels. I suggest, that is, that Holmes's twinned vocations as physician and creative writer impel his logic, bringing questions of biological reproduction and truthful representation together in such a way that they provide grounds for joining trauma and sexual inversion.

Trauma would be for Holmes a disorder of representation partly because (as with hysteria) its symptoms do not plainly lead to its causes—a dynamic clear enough in this novel from the extended puzzlement and ruminations of Lurida Vincent and Dr. Butts over Kirkwood's ailment. It would also disorder representation because the traumatic memory derails truthful representation of the present—as the frightful image from the past, in Kirkwood's case, now impedes the present apprehension of a beautiful woman. That is, trauma does not directly register itself, and in turn it deranges the body's natural process of registering and representing reality; the channels of expression and communication are perverted. In this case, as well, trauma as a malady of representation causes another malady of representation, sexual inversion. Obviously enough, as I have suggested, sexual inversion is a disorder of representation because it disarranges the natural outward expression of life, of life force, of sexual energy—and it of course foils reproduction. Indeed, both trauma and sexual inversion here impede hereditary transmission by disrupting the natural flow of human life-expression, of species self-representation. The anchoring particular that joins these ailments is an impediment or diversion in the expressive *flow* of life force. We ultimately see in *A Mortal Antipathy* a push toward a horizon of meaning where trauma and sexual inversion join as opposites of self-representation in both its communicative and biological senses. And this ground for their similarity—the way they work analogously, almost homologously, to impede natural expression—reappears in their cure, as I shall explain in a moment.

Kirkwood's cure intriguingly echoes the therapy of cathartically re-experiencing trauma that Freud and Breuer would later develop, which once again raises the question of where Holmes might have gotten the idea. Holmes's cure draws on several other strands of thinking about trauma and the nervous system, however. First, the report to the Royal Academy of Biological Sciences had suggested that Kirkwood's unnatural neural circuit could be "made to change its course, so as to follow its natural channel," possibly through "some sudden, unexpected impression," though the restoration of his "true nature" might require "as energetic a shock to displace it as it did to bring it into existence." For example, "If he were drowning, and a young woman should rescue him, it is by no means impossible that the change in the nervous current we have referred to might be brought about as rapidly, as easily, as the reversal of the poles in a magnet" (*MA*, 237–9). Such a cure involves a revision of the traumatic imprint, a re-righting of the perverted physical structure of neural pathways. And Kirkwood is indeed cured in a similar scenario. While sick, and asleep, he has a nightmare that he is deep in a coal mine, then that he is climbing a sheet of ice above an abyss, and then that the abyss is a volcano whose smoke is smothering him. At that, he awakens to realize the house is on fire. Paralyzed with fear, expecting death, he recalls his life-memories and, in particular, "The dread moment which had blighted his life returned in all its terror. He felt the convulsive spring in the form of a faint, impotent spasm,—the rush of air,—the thorns of the stinging and lacerating cradle into which he was precipitated" (*MA*, 263–6). This time, however, the narrative of the trauma is revised, as the heroine of the novel, Euthymia, rushes into the burning house and carries Kirkwood out "as easily as if he had been a babe." Outside, she holds him on her bosom, he hears her heart and feels her breathing, and she becomes associated with life rather than death (*MA*, 273–5). She causes "a revolution in his nervous system," his "vital currents" are redirected, and his "disturbed, perverted instincts" are rerouted into "their natural channel" (*MA*, 277, 281). The result: "The river which has found a new channel widens and deepens it; it lets the old watercourse fill up, and never returns to its forsaken bed. The tyrannous habit was broken" (*MA*, 279). The repeated story and its neural pathway have a new ending. Cradled like a babe, and reverting to his babyhood trauma, Kirkwood is reborn, freed from the image ingrained in the nerve fibers of his memory.

One might argue that, rather than a cathartic-reliving remedy, this cure more closely resembles the therapy of revising a traumatic memory through hypnotic suggestion, which Pierre Janet was soon to recommend. That is, in his thesis *L'automatisme psychologique* (1889), Janet, who believed in the psychogenesis of traumatic hysteria, describes cases in which neurotic patients who have suffered trauma are taken back through hypnotism to the moment of trauma, and then have the traumatic memory erased and replaced by hypnotic suggestion. For example, a patient who as a child saw an old woman fall down the stairs and die, and has hallucinatory relivings of the terror she felt, is taken back to the moment, and told that the old woman *did not* die. As Janet put it, after bringing the patient back "*to the moment of the accident*, I was able, not without difficulty, *to change the mental image*, to show her that the old woman had only stumbled and not killed herself, and efface the terrifying idea: the attacks of terror did not recur." He modifies or effaces the mental images from the moment of trauma, and thus accomplishes a cure by revising the patient's past.[30] Writing of the development of this process, Henri Ellenberger credits Janet with the first case of a cure through reliving a trauma—in work undertaken in 1888, and published in his thesis of 1889; Holmes's earlier fictional cure of course wouldn't count as a therapeutic breakthrough, but its precedence is astonishing. Malcolm Macmillan notes that the Belgian psychophysicist J.R.L. Delboeuf, also in 1888, used hypnotism in a very similar way to have a patient relive and expel a traumatic memory. According to Delboeuf, maladies can arise from a "suggestion" occurring at the traumatic moment; to effect a cure, the hypnotist "puts the subject back into the state in which his trouble manifested itself and combats with the spoken word the same trouble, but in a state of rebirth."[31] Both of these procedures assume that the patients, plunged into a traumatic state akin to the hypnotic state, had a fixed, pathological idea implanted in their memory; the cure was to use hypnotic suggestion to replace the idea.

Relatedly, the American author Edward Bellamy, who would become famous in 1888 for his novel *Looking Backward*, had published in 1880 a novel called *Dr. Heidenhoff's Process*, whose titular process was named "thought extirpation." On the principle that memories are stored in certain groups of "nerve corpuscles or fibres in the grey substance of the brain," and that these fibers and the memories they contain can be destroyed without affecting the rest of the brain, Dr. Heidenhoff applies galvanic current as a way of "obliterating the obnoxious train of recollections." Though he is able only to ablate morbid or diseased fibers, made so "by excessive indulgence of any particular train of thought," he foresees a day when "the mental physician will be able to extract a specific recollection from the memory as readily as a dentist pulls a tooth." The heroine of the novel, who wants to be freed from the memory and shame of having been seduced and abandoned, is told to "fix your attention on the class of memories which you wish destroyed; the electric current more readily follows the fibres which are being excited by the present passage of nerve force." The process works—until the surprise ending, when the hero awakes, realizes Dr. Heidenhoff's process was all a dream, and receives a letter from the heroine announcing that her shame will indeed be washed away, by her death, apparently by her own hand.[32] Bellamy's story preceded Holmes's novel, and Holmes undoubtedly knew it. But Heidenhoff's process has a similarity to Janet's and Delboeuf's which makes it fundamentally different from the process Maurice Kirkwood goes through: all these therapies provide cure or rebirth by forcing memory into misrepresentation, or by destroying truthful memory traces altogether.

As Ian Hacking has observed, Janet's therapy was not one of helping patients see the truth about their illnesses. The obscured, the hidden, was not revealed. Rather, the malleability of memory was exploited so that the image of the actual past was replaced by a curative fiction, a misrepresentation.[33] This decidedly is not the thrust of Holmes's cure. Rather, for Kirkwood, his "true" and "natural" self has been written over by his trauma. The natural expression of his life force has been impeded, or channeled off in the wrong direction. The cure is to enable the natural upsurge of vital force, an upsurge conflated with a sexual outflowing of nervous energy, culminating in biological reproduction. This release of vital nerve force is the true expression of Kirkwood's manhood and the vehicle for its reproduction. To drive this point finally home, Holmes has Kirkwood and Euthymia get married and produce a son, and Holmes ends his novel with a test. Partly curious to see if his antipathy to his cousin Laura is fully gone, but more interested to see if he may have passed his antipathy down to Maurice Jr., Kirkwood invites Laura to come see them. Happily, Laura elicits fear from neither father nor son (*MA*, 303–5). In the scientific world, this of course would be a repudiation of the still-pervasive Lamarckian notion that an acquired trait could be passed down to one's offspring. In Holmes's logic of expression, reproduction, and representation, it also means that, now that the perversion/impediment is removed, the hereditary sequence that naturally reproduces the image of the father in the son is restored, unmarked by any registration of trauma. The traumatic sexual inversion is cured by freeing the natural expression of life, by lifting the impediment, the nebula, to allow the outflowing of truthful human nature.

In *A Mortal Antipathy*, trauma and sexual inversion thus become a pair, imagined that way because they are maladies of representation, enemies of the true expression of life force. We see here an instance of literary imagining that foretells directions of scientific investigation—uncannily preceding Freud's 1890s work on sexual trauma and neurosis, the "antipathic instinct" of Krafft-Ebing's *Psychopathia Sexualis* (1886), the therapeutic re-experiencing of trauma that Janet and Freud would soon explore, and the much-later theorizing about homosexuality as caused by trauma. Holmes's precognition happens, I suggest, because this literary imagining crystallizes larger frameworks of meaning that can help us better understand and explain the trajectory of both scientific and cultural formulations and that can help us see, in this case, the ideas afloat—about neural force, reproduction, and species self-expression—that would lead to a pathologizing of sexual inversion. It helps us see how the very definition of sexual inversion would lead to its stigmatization as the opposite of healthy self-expression and reproduction. In short, we have a literary investigation that illuminates the long history of homosexuality pathologized as a wounding, and that foregrounds the formative assumptions that lay behind this conceptualization—assumptions that lurk behind similar pathologizations today. Biological theory here becomes intertwined with the theory of representation, and a science that pathologizes sexual inversion supports an emergent literary realism, which supports that science in return.

Notes

1. Oliver Wendell Holmes, *A Mortal Antipathy*, 3rd edn (Boston: Houghton, Mifflin, 1886). Further references will be cited parenthetically in the text as *MA*.
2. Sigmund Freud, "Three Essays on the Theory of Sexuality," in Peter Gay, ed., *The Freud Reader* (New York and London: W.W. Norton, 1989), 239–93, 241, 243–4.
3. Jane F. Thrailkill, in her recent book, *Affecting Fictions: Mind, Body, and Emotion in American Literary Realism* (Cambridge, Massachusetts: Harvard University Press, 2007), similarly situates *A Mortal Antipathy* in relation to the arguments about railway spine—as part of her general argument that we should renew our attention to the ways nineteenth-century writers conceived of embodied

emotion. Although the particulars of our arguments and contextualizations differ, we fundamentally agree that in his thinking about "shock," Holmes productively focused on the uncertain territory between corporeal and emotional etiologies.

4. See Ralph Harrington, "The Railway Accident: Trains, Trauma, and Technological Crisis in Nineteenth-Century Britain" (1999), http://www.york.ac.uk/inst/irs/irshome/papers/rlyacc.htm (accessed 24 January 2007), for a full history of railway spine and the explanations for it, from anatomical, to physiological, to psychological; Harrington gives special and full attention to John Eric Erichsen's and Herbert Page's developing theories. Also see Jill L. Matus, "Trauma, Memory, and Railway Disaster: The Dickensian Connection," *Victorian Studies* 43.3 (2001), 413–36 and Matus's essay in this volume. For a history of railway spine on the American scene, see Eric Caplan, *Mind Games: American Culture and the Birth of Psychotherapy* (Berkeley: University of California Press, 1998), 11–36. And for a more general account of the psychologization of trauma see Ian Hacking, *Rewriting the Soul: Multiple Personality and the Sciences of Memory* (Princeton, New Jersey: Princeton University Press, 1995), 183–96.
5. Herbert W. Page, *Injuries of the Spine and Spinal Cord Without Apparent Lesion, and Nervous Shock, in their Surgical and Medico-Legal Aspects* (London: J. & A. Churchill, 1883), 147–8; further references will be cited parenthetically in the text as *ISSC*. Holmes had also won the Boylston Prize, had served on the judging committee, and would undoubtedly have read Page's work. See Eleanor M. Tilton, *Amiable Autocrat: A Biography of Dr. Oliver Wendell Holmes* (New York: Henry Schuman, 1947), 140–2, 144–6, 185.
6. For a considerably earlier example, see Russell Reynolds, "Remarks on Paralysis and Other Disorders of Motion and Sensation, Dependent on Idea," *British Medical Journal* 2 (1869), 483–5, which argues that disorders of the nervous system could be caused by an emotion or an idea and the effects of shock after railway accidents are instances of this (484). Ralph Harrington notes that Frederic Le Gros Clark in 1868 and John Furneaux Jordan in 1873 also took psychological shock into account as a cause of railway-accident maladies. Arguing more generally for the effect of the mind on the body, Henry Maudsley in 1870 told the Royal College of Physicians that "A sudden and great mental shock may, like a great physical shock, and perhaps in the same way, paralyze for a time all the bodily and mental functions, or cause instant death" (*Body and Mind*, rev. edn [New York: Appleton, 1885], 38–9).
7. See James J. Putnam, "Recent Investigations into the Pathology of So-called Concussion of the Spine," *Boston Medical and Surgical Journal* 109 (1883), 217–20, which argues against the diagnosis of spinal concussion and for that of hysteria; J.J. Putnam, "The Medico-Legal Significance of Hemianaesthesia After Concussion Accidents," *American Journal of Neurology and Psychiatry* 3 (1884), 507–16, which focuses on male patients and suggests a diagnosis of "neurasthenia," partly because hysteria has come to suggest "exaggeration, womanishness, and deception" (512); G.L. Walton, "Case of Typical Hysterical Hemianesthesia in a Man Following Injury," *Archives of Medicine* 10 (1883), 88–95, largely about treatment with magnets, but also occupied with male subjects; and G.L. Walton, "Case of Hysterical Hemianaesthesia,

Convulsions and Motor Paralysis Brought on by a Fall," *Boston Medical and Surgical Journal* 111 (1884), 558–9, which stresses hysterical symptoms brought on by emotional trauma. Also see Caplan, *Mind Games*, 20.

8. Caplan, *Mind Games*, 21–3, and Mark S. Micale, "Charcot and the Idea of Hysteria in the Male: Gender, Mental Science, and Medical Diagnosis in Late Nineteenth-Century France," *Medical History* 34 (1990), 363–411, especially 370–1, 385–91, 406–8. As Micale observes, the extent to which Charcot progressed toward psychogenic explanations of hysteria is a topic of dispute, but while he clung to neurophysiological etiologies, he seems to have "moved timidly toward psychogenic explanations through his work on select topics." In several instances these topics included male hysteria (Micale, 391 n. 118). For additional histories of the development of the diagnosis of male hysteria, and Charcot's role in this, see Elaine Showalter, *Hystories: Hysterical Epidemics and Modern Culture* (New York: Columbia University Press, 1997), 62–77; Frank J. Sulloway, *Freud, Biologist of the Mind: Beyond the Psychoanalytic Legend* (New York: Basic Books, 1979), 30–41; Henri F. Ellenberger, *The Discovery of the Unconscious: The History and Evolution of Dynamic Psychiatry* (New York: Basic Books, 1970), 438–42; and Christopher G. Goetz, Michel Bonduelle, and Toby Gelfand, *Charcot: Constructing Neurology* (New York: Oxford University Press, 1995), 202–3.
9. See George Frederick Drinka, *The Birth of Neurosis: Myth, Malady, and the Victorians* (New York: Simon and Schuster, 1984), 114–6; and Michael Roth, "Hysterical Remembering," *Modernism/Modernity* 3.2 (1996), 1–30, 2–4. Henry Maudsley, in 1870, insisted on a similarly physical model of shock as causing "commotion in the molecules of the brain" (*Body and Mind*, 94). Also see Matus's essay in this volume for an excellent account of theories of nerve force as electrical and for another instance, in George Eliot's *Middlemarch* (1871–2), connecting mental shock to electrical shock.
10. Ellenberger, *The Discovery of the Unconscious*, 90–1, 148, 438–9; also Sulloway, *Freud, Biologist of the Mind*, 30, 33–4.
11. Roger Smith, *Inhibition: History and Meaning in the Sciences of Mind and Brain* (Berkeley: University of California Press, 1992), 75.
12. See John Furneaux Jordan, "On Shock After Surgical Operations and Injuries," *Special Enquiries; Including the Hastings Essay on Shock, the Treatment of Surgical Inflammations, and Numerous Clinical Lectures*, 2nd edn (London: J.A. Churchill, 1881), 1–60. Further references will be cited parenthetically in the text as *SASO*.
13. Smith, *Inhibition*, 76, 80–3.
14. On this development in the thought, especially, of Bain and Spencer, see Robert M. Young, *Mind, Brain and Adaptation in the Nineteenth Century: Cerebral Localization and its Biological Context from Gall to Ferrier* (New York: Oxford University Press, 1990), and Rick Rylance, *Victorian Psychology and British Culture 1850–1880* (New York: Oxford University Press, 2000). Also see Peter Amacher, "Freud's Neurological Education and its Influence on Psychoanalytic Theory," *Psychological Issues* 4.4 (1965), 21–8, on the brain anatomist Theodore Meynert's theories about the "arciform bundles" and "association fibres"—nerve bundles in the brain's white matter—as forming the basis for mental associations.
15. Théodule Ribot, *The Diseases of Memory* (New York: J. Fitzgerald, 1883), 2, 7.

16. F.G. Gosling, *Before Freud: Neurasthenia and the American Medical Community, 1870–1910* (Urbana and Chicago: University of Illinois Press, 1987), 132.
17. See, especially, George Chauncey Jr., "From Sexual Inversion to Homosexuality: The Changing Medical Conceptualization of Female 'Deviance,'" in Kathy Peiss and Christina Simmons, eds, *Passion and Power: Sexuality in History* (Philadelphia: Temple University Press, 1989), 87–117, and John Marshall, "Pansies, Perverts and Macho Men: Changing Conceptions of Male Homosexuality," in Kenneth Plummer, ed., *The Making of the Modern Homosexual* (Totowa, New Jersey: Barnes and Noble, 1981), 133–54. But also see Jeffrey Weeks, *Coming Out: Homosexual Politics in Britain, from the Nineteenth Century to the Present* (London: Quartet Books, 1977), especially the first two chapters; Neil Miller, *Out of the Past: Gay and Lesbian History from 1869 to the Present* (New York: Vintage, 1995), especially Chapter 2, "Pioneers of Sexology"; John D'Emilio and Estelle B. Freedman, *Intimate Matters: A History of Sexuality in America* (New York: Harper and Row, 1988), 121–2; Jonathan Ned Katz, *Gay/Lesbian Almanac: A New Documentary* (New York: Harper and Row, 1983), especially 139–57; and Francis Mark Mondimore, *A Natural History of Homosexuality* (Baltimore: Johns Hopkins University Press, 1996), 34–49.
18. Harry Oosterhuis, *Stepchildren of Nature: Krafft-Ebing, Psychiatry, and the Making of Sexual Identity* (Chicago: University of Chicago Press, 2000); Vernon A. Rosario, *The Erotic Imagination: French Histories of Perversity* (New York: Oxford University Press, 1997), 84. Of course, these scholars build on the history of the medicalization of homosexuality laid out by Michel Foucault in volume one of *The History of Sexuality*, 3 vols (New York: Random House, 1980), 1: 43.
19. The essays by Americans are: E.C. Spitzka, "A Historical Case of Sexual Perversion," *Chicago Medical Review* 4.4 (August 1881), 378–9; G. Alder Blumer, "A Case of Perverted Sexual Instinct," *American Journal of Insanity* 39 (July 1882), 22–35; James G. Kiernan, "Insanity. Lecture XXVI.—Sexual Perversion," *Detroit Lancet* 7.11 (May 1884), 481–4; James G. Kiernan, "Perverted Sexual Instinct," *Chicago Medical Journal and Examiner* 48 (1884), 263–5 (actually a report on a paper read by Kiernan); and J.C. Shaw and G.N. Ferris, "Perverted Sexual Instinct," *Journal of Nervous and Mental Disease* 10.2 (April 1883), 185–204. William M. Hammond, *Sexual Impotence in the Male* (New York: Birmingham, 1883) also devotes substantial attention to "perversion of the sexual appetite" and to pederasty as causes of heterosexual impotence.
20. See Shaw and Ferris, "Perverted Sexual Instinct," which helpfully summarizes previously reported cases.
21. Jean-Martin Charcot and Valentin Magnan, "Inversion du sens génital," *Archives de neurologie* 3 (1882), 53–60; and 4 (1882), 296–322.
22. Mark S. Micale plausibly suggests that Charcot, wanting to legitimize his diagnostic category of masculine hysteria, purposely kept it separate from homosexuality, so that male hysterics would not be marginalized as effeminate or as homosexual. Since, for Charcot, traumatic hysteria was so closely identified with male hysteria, this may have been the reason for not bringing trauma together with sexual inversion. See Micale, "Charcot and the Idea of Hysteria in the Male," 381–2.
23. Rosario, *The Erotic Imagination*, 85–9.

24. Ian Dowbiggin, "Back to the Future: Valentin Magnan, French Psychiatry, and the Classification of Mental Diseases, 1885–1925," *Social History of Medicine* 9.3 (1996), 383–409, 389–94.
25. Kiernan, "Insanity. Lecture XXVI.—Sexual Perversion," 481.
26. Blumer, "A Case of Perverted Sexual Instinct," 35.
27. Hammond, *Sexual Impotence*, 62–3, referring to a "decided inversion of the sexual feeling" in a patient who preferred the passive position in anal intercourse.
28. Shaw and Ferris, "Perverted Sexual Intinct," 190, 195, 199, 200.
29. Hammond, *Sexual Impotence*, 35–6.
30. Pierre Janet, *L'automatisme psychologique* (Paris: Alcan, 1889), 439. Ian Hacking provides a narrative of the psychologization of trauma that includes Janet's cases of reliving and revising traumatic memories and moves to Freud's (and Breuer's) method of the talking cure. Hacking also places Janet's work around 1888, and Freud, only in 1888, airs the idea of having a patient return in hypnosis to the traumatic event in order to cure it. Holmes's cure through reliving the trauma still predates both. See Hacking, *Rewriting the Soul*, 183–95.
31. Qtd. in Malcolm Macmillan, *Freud Evaluated: The Completed Arc* (Cambridge, Massachusetts: MIT Press, 1997), 77. As Macmillan reports, Delboeuf described his case in *Revue de Belgique* (15 November 1888)—with three succeeding parts, all published together as a book in 1889. Delboeuf's book, *Le magnetisme animal*, is reviewed in *Mind* 14 (1889), 470–1.
32. Edward Bellamy, *Dr. Heidenhoff's Process* (1880; New York: AMS Press, 1969), 96, 100–1, 110.
33. Hacking, *Rewriting the Soul*, 195–6.

6

Singing the Body Electric: Nervous Music and Sexuality in *Fin-de-Siècle* Literature

James Kennaway

Of all art forms, music has perhaps been the most closely associated with nerves. The very fact that music has only limited powers of representation has meant that descriptions of its effects have often turned less to its "content" and more to its physical impact on the body. And throughout the history of neurology nerves themselves have been compared to the tightened strings of a musical instrument. George Cheyne's *The English Malady* from 1733, for instance, used a musical analogy for the whole nervous system, writing of "the Brain, where all the nerves, or instruments of Sensation terminate, like a Musician in a finely fram'd and well-tun'd Organ-Case . . . these nerves are like Keys, which, being struck or touch'd, convey the Sound and Harmony to this sentient Principle, or Musician."[1] As we shall see, the relationship between music and nerves went far beyond such metaphors, and by the end of the nineteenth century had become one of the most important ways of talking about music. The idea of musical over-stimulation of the nervous system was the basis of a whole discourse in literature, music criticism and psychiatry that saw music as a potential threat. This chapter seeks to demonstrate that the debate on nervous music was intimately connected to broader cultural anxieties about sexuality, and the way it related to personal autonomy and social order.[2] The nineteenth-century "body electric," to borrow Walt Whitman's term, was not the automaton of the twentieth-century imagination, but a body of sexualised-electric nerves.[3]

From the 1880s a potent mixture of a bourgeois scientific fear of the medical and moral consequences of certain music and a Decadent appreciation of that music created a rich and contradictory literature about music and nerves. In works by Thomas Mann, George Gissing, George Du Maurier and many less well-known figures, music, especially that of Richard Wagner, crops up again and again as a real threat to health. Unlikely as it may sound, characters regularly lose self-control, have breakdowns and sometimes die from listening to or performing particular kinds of music. This music, it was believed, could wear out, and indeed overwhelm, listeners' electric nervous systems. In contrast to our own era, what the nineteenth century considered "electric music" was not viewed as machine-like and objective, but as essentially subjective and sensual.[4]

New scientific understanding of the nerves laid the groundwork for literary discourse on nervous music. Luigi Galvani's discovery of animal electricity in 1789 had profound consequences both for science and for cultural attitudes towards the body. The idea of nerves as essentially electrical was reinforced when the discoveries of Hermann Helmholtz, Emil Du Bois-Reymond, Guillaume Duchenne, and the brain stimulation experiments by Hitzig and Fritsch firmly established the electrical basis of the nervous system. The idea of music as a business of electricity and nerves received a boost from Helmholtz's experiments of 1847 which measured the speed of nerve conduction and equated it with electricity. Thereafter, the particularly "nervous" effects of music became a commonplace in science and literature.

This was an era in which electrotherapy boomed for a wide variety of maladies, often sexual in character. Treatments ranged from the Herculex belt, an electric belt on sale on the open market,[5] to the electrocution of patients' genitals – striking evidence of the relationship between nerves and sexuality in contemporary discourse.[6] Erotic strain on these electric nerves was seen as an important cause of physical and psychiatric conditions. Sensuality (that favorite Victorian euphemism for eroticism) was regarded as being both morally suspect and bad for the nerves in general. The novels of the turn of the century provide many examples that show that this was equally true for musical sensuality.

Those with so-called weak nerves, notably, as we shall see, women and homosexuals, were especially at risk from sensual music.[7] The famous sexologist Richard Freiherr von Krafft-Ebing was one of many who argued that such music could cause actual neurotic conditions:

> One of the monstrosities of our modern civilization is the idea that every child of the higher classes must have a musical education. No one asks about talent. Fashion demands that young people can tinkle on the piano . . . In cases of limited talent studying the piano is an inadequate achievement which makes heavy demands on the physical and mental strength of the player and often creates nervousness and in the case of untalented girls, if practiced too

intensively or unwillingly, can become the cause of serious nervous diseases.[8]

The nervous diseases in question included a wide range of sexual and psychiatric conditions, and the strain on the nerves of listening to music was thus seen as a real medical issue. Music, like sex, was a stimulant of the nerves, unhealthy in excess, and implying an alarming loss of self-control. In the literature of the *fin de siècle,* female characters often lose their morals and their health from contact with such musical contagion.

These vulnerable nerves became the paradigm in which the nineteenth century discussed the age-old theme of the perils of musical sensuality. The supposed medical dangers of music reflected moral and political anxieties about music that have been a regular feature of Western musical history.[9] At least since Damon of Oa, a contemporary of Socrates, commentators have argued that music is essentially a question of order and not of pleasure. Music was there to form willpower and masculine character in the interest of society, not to indulge an individual's senses. It was on these grounds that Plato wanted to ban music that made men effeminate and sensual, allowing only the Phrygian and Dorian modes.[10] The nineteenth-century debate on nervous music is in some ways a continuation of this theme with scientific vocabulary. However, the model of nervous stimulation importantly shifted the emphasis from ethics to biology, giving moral scruples some of the prestige of nineteenth-century science. Music's potential power to subvert self-control, notably in sexual matters, via the nerves made it not only a danger to individual health but possibly also to society. Some cultural commentators, writers, and psychiatrists argued that music could be a cause or a symptom of mass neurasthenia and broader cultural malaise.

The centrality of the nerves to ideas of the power of music, and to health and sickness, goes back to the scientific revolution of the seventeenth century. Although neo-Platonism still had considerable influence on what came to be known as the *Affektenlehre,* which explained the effects of music with objective correspondences between keys and states of mind rather than empirical evidence of the function of the auditory nerves, the physical aspect of music was becoming more emphasised. Descartes's mechanistic view of the world put more focus on the physical effect of sound on the body via "animal spirits" in the nerves. At the same time, the work of Thomas Willis and those that followed him increased knowledge of the physical nature of the nervous system. Along with the analogy with rhetoric and the physics of acoustics, the effect of music on the nerves was thus a key aspect of discourse on music in the seventeenth and eighteenth centuries.

As the nerves became more important in discussions of music, they also came to be central in medical diagnoses of physical and mental illness. Drawing on Francis Glisson and Albrecht von Haller's work on the irritability of nerves, Scottish Enlightenment figures such as John Brown and George Cheyne created what might be called the nerve paradigm of disease, at times asserting that nerves were responsible for most sickness, both mental and physical. Indeed, from the work of William Cullen (who coined the neologism "neurosis" in 1769) until Freud, the aetiology of psychiatric conditions was widely believed to be in some sense a question of the nerves. Health was thus a question of regimen, of maintaining a balance in life and avoiding over-stimulation of the nerves through excess of any kind.

During the nineteenth century, the associations between music and nerves and music and illness came together, creating a consensus that music could exhaust the nerves and cause actual "nervous" (that is, neurotic and psychiatric) symptoms. In *Physiological Aesthetics* (1877), the Canadian scientist and novelist Grant Allen summed up contemporary thinking on the subject, writing that the "auditory nerves are not liable to be scratched, burned, bruised, or attacked by chemical agents, but only to be wearied by over-use or jarred by discordant sounds."[11] Increased understanding of neurology appeared to reveal a direct, electrical link between sounds and consciousness, between physiology and psychology. It was precisely this assumption that music could tire out the nerves that connected music directly to the discourse of neurasthenia and the social, cultural, and sexual issues it raised. As listening, and consciousness generally, came to be viewed more physiologically, discourse on music in general became "scientific" and fixated on the body – a body understood primarily in medical and erotic terms.

The idea of music as a potential pathogen had its Golden Age in the final decades of the nineteenth century. George Beard's neurasthenia diagnosis from 1869 created a potent combination of the doctrine of nervous stimulation, the understanding of the electrical character of the nervous system, and a sexual aetiology. The diagnosis put tired, over-stimulated nerves at the heart of psychiatry. In a parallel development, the establishment of degeneration theory in the 1850s, which saw bad heredity as the principal cause of mental illness, meant that music's power over the nervous system was increasingly seen in the context of hereditary nervous weakness. In contrast to the eighteenth century, when the effect of music on the nerves had been seen more in the context of a regimen of self-control leading to moral improvement, many in the late-nineteenth century saw biology as destiny, and music as a threat to more than individual well-being. Because of this, nervousness, and therefore also nervous music, began to lose some of the elite glamour it had previously enjoyed. Increasingly it was seen as a mass phenomenon, even as a threat to civilisation, as was reflected in Friedrich Nietzsche's critique of Wagner from the 1880s and Max Nordau's famous book *Entartung* from 1892.

Dangerous erotic music and nerves in *fin-de-siècle* literature

At the same time as these jeremiads were lamenting the effects of nervous music, those who felt alienated by Victorian notions of health and progress turned to the ideas of nervousness and decadence as alternatives. The great forebear of *fin-de-siècle* Decadence, Charles Baudelaire, wrote approvingly of the "intensité nerveuse" of music in his influential 1861 article *Richard Wagner et Tannhaeuser in Paris.*[12] The Decadents of the 1880s and 1890s followed Baudelaire in this attitude, especially with regard to Richard Wagner. *La Revue wagnérienne* (1885–8) was a periodical dedicated to spreading Wagner's works and artistic theories that could count many prominent Decadents among its contributors. From this point on there was a constant stream of novels, stories and poems about Wagner, and nervous music in general.[13] These Decadents were drawn to Wagner by the urge to *épater le bourgeois* by praising a notorious hater of France, but especially by the erotic overtones and illicit themes in his work.[14]

The over-the-top French Decadent prose of writers such as Joséphin Péladan, Catulle Mendes and Élémir Bourges, in which nervous music symbolises erotic freedom and a rejection of contemporary values, is one of the two main ways in which the theme of nervous music arises in the novels of the period. In these books, the extreme effects on the nerves of listening to music are welcomed even though they can bring death. They see the kind of nervous breakdown brought on by the strain of hearing music as a transfiguration, a glorification of illness that is such a marked characteristic of the Decadents. In this version, suffering from music is shown to be a sign of having a higher nature, above mere bourgeois health – sometimes as an overt elaboration of the Wagnerian theme of the love-death.

The second version of this theme is more straightforwardly critical of music, which is seen as a potential pathogen to be avoided for the sake of mental and physical health. In novels like George Du Maurier's *Trilby* (1894) or George Gissing's *The Whirlpool* (1897), music can be both a symptom and a cause of nervousness. Its victims are either innocents corrupted by the wicked or else themselves already marked by the stigmata of degeneration. The sexual freedom that the Decadents saw in nervous music is a threat to self-control and social order, not a welcome escape from them. This mode of talking about nerves and music was especially common in England (notorious as the "Land ohne Musik"), and also in the United States.

Although they may disagree on how to judge nervous music, both attitudes share an understanding of what sort of music is involved. It is clear that certain kinds of music are regarded as healthy and others are seen as especially nervous. Over and over again the same names appear: Frédéric Chopin, Peter Illyich Tchaikovsky, Franz Schubert, but above all Richard Wagner. To a certain extent, biographical factors help explain this; Chopin's famous tuberculosis, Tchaikovsky's homosexuality and tragic death, a certain morbid tendency in Schubert, all these things helped mark them out for this kind of treatment. However, I argue that musical features also play a key role. The music of the so-called *neudeutsche Schule* of Wagner, Liszt and Berlioz was regularly attacked (and occasionally praised) for being more "nervous" than the more conservative style of Brahms or Mendelssohn.[15]

Bearing this in mind, it is perhaps possible to point to specific musical characteristics that provoked accusations that a composer's work caused nervousness. Above all, it seems that what was heard as an erotic musical rhetoric was at the heart of the issue. The music that is regularly described as nervous tends to have more daring harmonies, a looser structure, perhaps drawn from literature, and lush instrumentation. In short, it was regarded as having elements that appealed to the ear, to the senses, and not to the rational mind. Even when the music was purely instrumental, it was often regarded as in some sense erotic. The specific musical elements that were regarded as most stimulating to the nerves were those that appealed to the senses, titillating the audience and undermining the rational disinterested enjoyment of music with dangerous, superficial pleasure.

This is particularly clear if one looks at what kinds of music were described as "electric." Alan Walker's biography of Franz Liszt records an occasion on which he was asked to play to a mental patient in an asylum: "The passage he played produced a visible effect on her similar to that of an electrical discharge."[16] Heinrich Heine's comments on Liszt use the language of erotic electricity. The virtuoso is described as demonstrating "magnetism, galvanism, electricity . . . histrionic epilepsy . . . the phenomenon of titillation . . . musical Spanish fly . . . and other scabrous things that have reference to the mysteries of the bona dea."[17] In a similar vein, the *Revue trimestrielle* in a cartoon portrayed Berlioz and Wagner in a mesmerizing competition, each trying to defeat the other electrically.[18] Others also saw the effect of Wagner's music as in some way sexual and electrical. In his *Geschichte des Dramas* (1871), J.L. Klein wrote about female Wagnerians in terms of Luigi Galvani's famous experiment of electrocuting frogs' legs: "the blasé hysterical female court parasites who need this galvanic [i.e. electrical] stimulation by massive instrumental treatment to throw their pleasure-weary frogs-legs into violent convulsions."[19] Again and again, music of this kind is portrayed as a form of electrical stimulation of the nerves, mostly with explicit sexual overtones.

Not everyone used electrical rhetoric to discuss Wagner's effect on the nerves, but the physiological and erotic power of music was one thing on which Wagnerians and anti-Wagnerians agreed. The death of Wagner's first Tristan, Ludwig Schnorr von Carolsfeld, in a Wagnerian delirium was only the most famous example of people "dying of *Tristan*." Wagner's diary shows that he felt that the music was partly to blame, writing of the singer that, "I drove you to the abyss," and many in the press connected his death to the nervous strain of Wagner's music.[20] Passing remarks by critics aside, discussion of the physical impact of Wagner began with the publication of Theodor Puschmann's *Richard Wagner – eine psychiatrische Studie* in 1872, where he outlined what he considered to be the composer's psychiatric symptoms. Puschmann explicitly links the sexual scandals in Wagner's private life, "his inversion of appetites, perversion of desires and wishes and in the total lack of moral and social feelings,"[21] to his "brain-shaking instrumentation" that requires, he says, an "aural nerve as thick as a ship's towing line."[22] In the journalism, medical literature, and novels of the period, the physical effect of listening to the advanced chromatic harmonies and loose forms of Wagner's music was a common theme, sometimes portrayed positively, sometimes negatively, but always as a potential threat to health and sanity.

Thomas Mann's 1903 story "Tristan" is a clear literary example of the subject of Wagner's music as pathogen. Mann's story uses his characteristic irony to satirise the Wagnerian love-death theme. At the same time, his deliberately banal and bourgeois version of the Tristan myth is typical for the era in that it sees nerves as the means by which the fatal erotic power of music works on its victims. It is set in a sanatorium called Einfried, an obvious reference to Wagner's house in Bayreuth, Wahnfried. The story concerns the encounter between Detlev Spinell, a writer, and

Gabriele Klöterjahn, the wife of a Baltic merchant. Spinell is not himself ill. He explains why he is at the sanatorium to Frau Klöterjahn: "Cure? Oh, I'm having myself electrified a bit. Nothing worth mentioning. I will tell you the real reason why I am here. It is a feeling for style."[23] Frau Klöterjahn, on the other hand, has been brought to the sanatorium by her extravagantly healthy husband because of a touch of tuberculosis.

The key moment of the story comes when Spinell persuades Gabriele Klöterjahn to play the piano. Her first reaction to the suggestion is a telling piece of evidence for the way music was understood at the time: "But our family physician, as well as Dr. Leander, expressly forbade it, Herr Spinell."[24] Gradually, however, she consents to play, and the repertoire involved is exactly what one would expect. First, she plays Nocturnes (opus 9, no. 2 and two others) by the famous tuberculosis victim, Frédéric Chopin. While playing, Mann writes, "She displayed a nervous feeling for modulations of timbre."[25] But Chopin is only the beginning. The real musical poison comes in the form of the score for Wagner's *Tristan und Isolde*. Gabriele Klöterjahn appears shocked to find such a piece in a sanatorium, as if it were obviously unsuitable. When shown the score by Spinell she says, "Really? How did that get here? Give it me."[26] The highpoint of her ecstatic music making comes during the second act, the famous prolonged love duet, with its radical harmonic tension. The nervous excitement is too much for Frau Klöterjahn. Two days later, what had been just a touch of infection turns into full-blown tuberculosis. Detlev Spinell writes a vindictive, snobbish letter to Gabriele Klöterjahn's husband, and as they argue over it, they are told that she has died after coughing up a great deal of blood.

Discussions of Wagner's nervous music often allude directly to its erotic character. The American novelist Gertrude Atherton in her book *Tower of Ivory* (1910) describes the effects on the soprano Styr of singing the role of Isolde in "the most licentious opera ever written."[27] After singing one evening, "Styr . . . wondered if she were alive or a walking automaton. Her passion had expended itself, the blood had left her brain."[28] The obvious background to this physiological language is sexual, as this extract of a letter written by Styr demonstrates:

> Dear Mr. Ordham,
> You will recall that I told you it would be better to think of me as a stage woman only? . . . If I cannot make you understand the fearful state of excitement which an opera like Isolde induces, then indeed I hope you will not forgive me, never come near me again. But I fancy you have already forgiven me. I was a wild beast.[29]

The book contains many references to the power of music on the nerves; Mrs. Cutting states that after seeing Wagner's *Die Walküre* "It was quite a week before I felt rested."[30] If any doubt remained about the decadence involved, the fact that Ordham and Styr discuss *À rebours* puts it to rest.[31] The consequences of submitting to the power of erotic nervous music are familiar: "No one was surprised to hear of his illness – brain fever?"[32] Over-stimulation, musically and sexually, could be fatal. Nervous music came to stand for transgressive sexuality in novels of the turn of the century, and such breaking of taboos was punished.

George Moore's *Evelyn Innes* from 1898 presents a similar tale of the dangers of nervous music. The eponymous heroine at one point says "to sing Isolde and live a chaste life, she did not believe it to be possible."[33] Her portrayal of Isolde on stage is reflection of her erotic life: "After the love potion had been drunk, the moment of her life to put on the stage was its moment of highest sexual exaltation."[34] After an upbringing surrounded by that epitome of musical virtue, Johann Sebastian Bach, Evelyn Innes is corrupted by Sir Owen Asher, a wealthy Wagnerite. He aims to turn her into a great soprano, as Svengali did with Trilby. Later, when Evelyn and Owen argue about why she became the lover of the composer Ulick Dean, Asher blames the music: "I can understand; all those hours you spent with him studying Isolde." To which she replies, "Yes, it was the music, it gets on one's nerves . . ."[35] Indeed, she confesses to Monsignor Mostyn that the music leads to such things: "But one night at the theatre, during the performance of *Tristan and Isolde,* I sinned with a second man."[36] Evelyn Innes does not die from the erotic nervous strain caused by the music, but becomes a nun.

One aspect of perceived erotic threat posed by Wagner's music that markedly contrasts with his macho reputation since the 1940s was the specter of homosexuality. Contemporaries who saw homosexuality as the emasculating result of a lack of willpower could regard nervous music, with its sensuality and hypnotic power, as a possible symptom, or even a cause, of effeminizing illicit sexuality.[37] The same musical features that made Wagner's music an apparent danger to female listeners and players (such as Evelyn Innes and Gabriele Klöterjahn), also associated it with homosexuality, since gay men, it was argued, had "weak" nerves like women. The sexologist Krafft-Ebing came across a number of gay Wagnerians in his extensive research. One homosexual interviewee said of Wagner: "I find that this music is perfectly in tune with our natures."[38] He went on to say:

> Among poets and novelists, I typically prefer those who describe refined feelings, peculiar passions, and far-fetched impressions; an

artificial or hyper-artificial style pleases me. Likewise in music, it is the nervous, exciting music of a Chopin, a Schumann, a Schubert, or a Wagner, etc. that is in most perfect harmony with me. Everything in art that is not only original, but also bizarre, attracts me.[39]

Thomas Mann's short story *Tonio Kröger* even seems to suggest that Wagner could turn sensitive young men gay. The protagonist refers to *Tristan and Isolde* as "a morbid and profoundly equivocal work" and warns of "the effect it has on a healthy young man of thoroughly normal feelings . . . I've seen women and youths go mad over artists."[40] In a similar vein, the French writer Catulle Mendes, who had known Wagner personally and written for *La Revue wagnérienne,* produced a very peculiar novel in 1880 entitled *Le Roi vierge* describing a fictionalized and rather homoerotic version of the relationship between Richard Wagner ("Hans Hammer'") and King Ludwig II of Bavaria ("Friedrich II of Thuringia"), who comes to a sticky end after his encounter with the composer.[41]

Wagner's reputation as a composer for homosexuals was partly based on the supposed effeminate luxury of his music. This association between Wagner and sensuality and luxury, both musical and personal, was set in the public imagination of the late-nineteenth century by the publication of his *Briefe an eine Putzmacherin* (Letters to a Seamstress). This was the correspondence between Wagner and the seamstress Bertha Goldwag. The letters were published after they came into the hands of Daniel Spitzer, a prominent Viennese journalist and writer.[42] He explicitly links luxury to the idea of nerves and degeneration, taking at times a deliberately medical tone:

> The writings of Lombroso, Krafft-Ebing and Nordau have become commonplace; helping us understand the relationship between the physical nature of man and his mental creations, binding them together into an indissoluble unity. Impartial biographies that understand the unity of psyche and body, show the value of these orders of satin.[43]

Daniel Spitzer later went as far as to write an entire novel mocking Wagner's preference for soft furnishings. *Verliebte Wagnerianer* ("Wagnerians in love") from 1880 is remarkable for the way in which it sticks to its subject. In one scene, Dr. Brauser explains the value of his and Wagner's contribution to music:

> Shallow heads, cynics and jokers mocked when it became known that Richard Wagner wore little satin pants, dressing gowns and little

> boots made of this material [satin] and changed the colour according to the composition he was working on. However he believed that if Wagner had never rhymed a line and never written a note and had achieved nothing but show by example to science the influence of satin on musical creativity, and this alone would have secured the immortality of his name. It is clear that satin was an essential for every composition, more reliable as a stimulant than coffee, tea or tobacco, indeed if the right experiments were conducted a satin-narcotic effect could be achieved.[44]

This extract also points to the way Wagner's music was perceived. It clearly implies effeminacy expressed in a quasi-narcotic stimulation of the nerves. Musical sensuality, condemned since Plato, could undermine masculinity by seducing the nerves.

Degenerate music and nerves

An eminently significant literary character who links homosexuality and the power of music on the nerves is Joris-Karl Huysmans' Duc Jean Floressas des Esseintes. The protagonist of the 1884 novel *À rebours* is a victim of the power of music and weakened heredity. Huysmans had contributed to *La Revue wagnérienne* and his protagonist may not be as well-versed in music as in other arts, but it too holds unhealthy thrills for him.

> He therefore resolutely abstained from all musical indulgence, and the only pleasant memories he retained from these years of abstinence were of certain chamber concerts at which he had heard some Beethoven and above all some Schumann and Schubert, which had stimulated his nerves in the same way as Poe's most intimate and anguished poems. Certain settings for the violincello by Schumann had left him panting with emotion, choking with hysteria; but it was chiefly Schubert's *Lieder* that excited him, carried him away, then prostrated him as if he had been squandering his nervous energy, indulging in a mystical debauch.
>
> He had never been able to hum *Des Mädchens Klage* without nervous tears rising to his eyes.[45]

Des Esseintes's homosexual tendency, his effeminacy, and his vulnerability are all due to his weakened heredity, to his status as a degenerate. Huysmans turns explicitly and repeatedly to the language of nerves to describe its effect on this product of an "effete, enfeebled race." The idea of degeneration owed its origins to the work of Benedict Morel, particularly his 1857 book *Traité des degenerescences physiques, intellectuelles et morales de l'espèce humaine et des causes qui produisant ces variétés maladives.* Drawing on Lamarck's theory of the inheritance of acquired characteristics, a pre-Darwinian conception of evolution, Morel argued that mankind could degenerate as easily as evolve, and that things like alcoholism and cretinism would lead to generations of increasingly atavistic specimens. From the time of its publication until thc 1940s, the theory proved very influential and left literary traces, most notably in Zola's Rougon-Macquart series. The perverse sexual nature of a des Esseintes, Zola's Nana, or those affected by nervous music was seen by many in the context of degeneration.

Other novels of the period also reflect this anxiety about the role of music in degeneration. George Gissing's *The Whirlpool*, an account of the marriage between Harvey Rolfe and Alma Frothringham, is one good example. Rolfe is the epitome of 1890s rugged manhood and makes approving comments about the wholesome barbarism of the British Empire and country life as cures for modern over-civilisation and nervousness. Alma, on the other hand, is a talented musician whose heredity is stained with degeneration. Her father dies of a morphia overdose, and since Morel's work, intoxication with drugs had been considered a sure way to create degenerate descendents. Her hereditary nervous weakness is linked to music as well as the accusations of adultery that lead to the novel's tragic denouement. The character of Rolfe is as interested in degeneration as Gissing himself. At one point he reads Ribot's treatise *L'hérédité psychologique* and comes to see in his wife's behaviour the sure marks of bad heredity.[46] "He knew Alma's characteristic weakness, and did not like to be so strongly reminded of it. For about this time he was reading and musing much on questions of heredity."[47]

The principal symptom of Alma's degenerate nature is her nervous relationship with music. There is scarcely a reference to her playing music that does not also bring in the idea of nerves. That this innate nervousness is part of the "artistic temperament" is clear from comments made by the character Mrs. Strangeways to Alma: "I am sure it must be very exhausting to the nerves; you have so undeniably the glow, the fervour of the artist."[48] Nor is this nervousness simply stage fright or high spirits – it is a symptom. Music brings out her innate nervous weakness. Before and after her first big public concert as a violinist, she suffers from neurological conditions. At one point she is unable to steady her hand, which she describes as "a neuralgic something . . . It's the price one pays for civilization."[49] Her husband describes her collapse after the concert with the words; "Oh, the nerves – breakdown – nothing dangerous, I believe."[50] As a woman with degenerate blood, Alma is doubly at risk from nervous music, and her early demise is testimony to its power over the body.

The most famous book to deal with music and degeneration, however, is certainly Thomas Mann's *Buddenbrooks* (1901). The subtitle of the novel, "Decline of a Family," alludes to its close links to ideas of hereditary degeneration. The representative of the final generation is Hanno Buddenbrook, more imaginative and more fragile than his ancestors. Hanno may perhaps be too young to die from music's erotic power in a direct sense, but in terms of his tainted heredity, he is the victim of a move from sturdy masculinity to feminine, artistic sensuality in his family. He is a nervous, highly strung child and when he listens to his mother playing Wagner on the piano, his teacher is furious:

> I cannot play it, my dear lady! I am your most devoted servant – but I cannot. That is not music – believe me! I have always flattered myself that I know something about music – but this is chaos! That is demagogy, blasphemy, madness! That is a perfumed fog shot through with lightning! That is the end of all honesty in art! I will not play it! . . . Look the child sits listening. Would you then utterly corrupt his soul?[51]

At first, the music teacher Pfühl is shown to be a lover of Bach and counterpoint – the opposite of Wagner's sensuality, and Hanno's mother's love of Wagner and new music is a sure sign of the degeneration of the family. Later Hanno uses Wagnerian musical tricks himself. Pfühl is again furious when he hears Wagnerian style modern harmony in Hanno's compositions:

> What sort of theatrical ending is that, Johann? That doesn't go with the rest of it. In the beginning it is all pretty good; but why do you suddenly fall from B flat major into the six-four chord on the fourth note with a minor third? These are tricks; and your tremolo here, too. Where did you pick that up? I know, of course; you have been listening when I played certain things for your mother.[52]

At the finale of *Buddenbrooks*, Hanno plays by himself and is soon struck by a nervous lethargy:

> Hanno sat still for a moment, his chin on his breast, his hands in his lap. Then he got up and closed the instrument. He was very pale, there was no strength in his knees, and his eyes were burning.

> He went into the next room, stretched himself out on the chaise-lounge and remained there for a long time motionless.[53]

At the start of the following chapter the reader learns that Hanno has been taken ill. Although he dies of typhoid (which is an infectious disease with little to do with the nerves), it is clear that Hanno dies of nervous music, and of a failure of the will to live that Mann, following Friedrich Nietzsche, saw in Wagner and Schopenhauer. A line from earlier in the novel could serve as the moral of the tale – "It is the nerves, Herr Senator; the nerves are to blame for everything."[54]

The dangers of musical hypnotism

Sensual, nervous music of the kind that leads to Hanno's early death could overwhelm the listener in a way that was often compared with drugs, and especially with hypnosis. Like hypnosis, music could undermine rational self-control and bring out latent desires. The connection between hypnosis and hysteria was given particular prominence by Charcot's public displays of the power of hypnosis on hysterics at the Salpêtrière, which themselves had a very strong (if unacknowledged) theatrical character. The American writer and critic James Huneker saw a parallel between Wagner's power over his audience and Charcot's hypnotic power over hysterics, writing that "this Klingsor of Bayreuth hypnotizes his hearers with two or three themes not of themselves remarkable, as Charcot controls his patients with a shining mirror."[55] Similarly, Leo Tolstoy's novella *The Kreutzer Sonata* contains a clear warning of the sexual dangers of musical mesmerism.

> Hence the terrible effects that music occasionally produces. In China music is a state concern, and this is as it ought to be. Could it be tolerated in any country that any one who takes the fancy may hypnotize any one else and then do with him whatever he has a mind to, especially if this magnetiseur is – Heaven knows who! – an immoral character, for instance?[56]

He portrays music as a kind of sinister mind-control that could rob the listener of self-control and lead to sexual immorality, two recurring themes in the debate on nervous music.

However, by far the most famous example of the sexual power of hypnotic music can be found in George Du Maurier's novel *Trilby*.[57] In it, the charismatic Jewish impresario, Svengali, hypnotizes the innocent girl Trilby, and uses her talent to tour Europe. In the case of *Trilby*, the hypnotism is quite literal: "Svengali told her to sit down on the divan and sat opposite to her, and bade her look him well in the white of the eyes."[58] The sexual and musical power of such hypnosis is demonstrated by the fact that the hideous Svengali marries the beautiful Trilby and that her voice attains mysterious power and expression. The question of effeminacy is also raised, as we read that Svengali's voice "often broke into a disagreeable falsetto."[59] The fact that Svengali is portrayed as "small and weak and flurried . . . an Oriental Israelite Hebrew Jew,"[60] raises many of the issues that Sander Gilman and others have addressed in the context of masculinity, Jewishness and nervousness.[61]

Comparisons between hypnosis and the effects of nervous music point to an essential characteristic of the idea of nerves in the *fin de siècle* – its supposedly subconscious, purely physical and therefore dangerously undisciplined nature. The supposed sexual danger associated with such music was directly related to its ability to overcome the listener's self-control. The direct stimulation of the body via nerves was contrasted with conscious subjectivity. For example, the novelist and critic Hermann Bahr, drawing on the work of Ernst Mach, argued that his contemporaries had a weakened sense of self and existed only through nerves: "They are nervous; the rest is extinct, withered and barren. They live now only through the experience of nerves; they only react on the basis of nerves."[62] The nerves, as the opposite of the conscious mind, could thus be seen as a source of imagination and creativity, but were more often marked as feminine and effeminate, a threat to masculine self-control and order.

Strikingly, the musical features usually described as nervous are so-called surface characteristics, such as timbre. The sensual pleasures of *Klang* (sound) offered none of the Kantian disinterested appreciation of art that Idealist aesthetics privileged: they were mere appearance. "Deep" structural features (especially form) and music that seemed to emphasise them were regarded as marks of healthy subjectivity. For hostile critics, Wagnerian chromatic harmony was more of a question of surface than of structure. Form was often described as an expression of the will, that is, as a model of subjectivity, and music that appeared to lack form was therefore associated with the tropes of weak will – hysteria, effeminacy, and the feminine. One sign of this association is the extent to which Wagnerians have attempted to establish that the Meister's works were constructed on formal principles and were not mere surface phenomena, and the way that Tchaikovsky and Schubert's "failure" with large Beethovenian forms was put in terms of questionable masculinity.[63]

Nervous music seemed to offer immediate sensual pleasure (as opposed to delayed gratification) and to undermine the will. The literary theme of nervous music therefore reflected a division in which certain kinds of music were relegated to a pseudo-medical discourse of neural over-stimulation. Other kinds of music maintained the status accorded to them by Idealist aesthetics and functioned in novels of the period as a sign of bourgeois high culture or a symbol of the quasi-religious sublime. All music was effectively divided up along these lines in a hierarchy of mind/body, masculine/feminine. By the *fin de siècle*, therefore, neurology had become one of the most important contexts for literary understandings of music and also a reflection of contemporary anxieties about masculinity, femininity, rationality, and the dangers of the imagination. Nervous music epitomised everything that seemed to contradict music's role as a model of cosmic and social order, which from Plato to attacks on "musical Bolshevism" and rock music in the twentieth century has been behind many of the moral panics around music. The idea of nerves gave a medical veneer to this perennial debate from the mid-nineteenth century to the heyday of the concept of degenerate music in Nazi Germany.

Notes

1. George Cheyne, *The English Malady* (London: G. Strahan, 1733), 4.
2. The last few years have seen a marked rise in interest in the subject of music and medicine. By dealing with music and nerves, this chapter follows the work in the history of medicine, musicology and literary studies on degeneration and culture. See Peregrine Horden, ed., *Music as Medicine: The History of Music Therapy since Antiquity* (Aldershot: Ashgate, 2000); Daniel Pick, *Svengali's Web: The Alien Enchanter in Modern Culture* (New Haven: Yale University Press, 2000); Alisa Clapp-Itnyre, *Angelic Airs, Subversive Songs: Music as Social Discourse in the Victorian Novel* (Athens: Ohio University Press, 2002); Phyllis Weliver, *Women Musicians in Victorian Fiction 1860–1900: Representations of Music, Science and Gender in the Leisured Home* (Aldershot: Ashgate, 2000); Sophie Fuller, *The Idea of Music in Victorian Fiction* (Aldershot: Ashgate, 2004); Emma Sutton, *Aubrey Beardsley and British Wagnerism in the 1890s* (Oxford: Oxford University Press, 2002); Sander Gilman, "Strauss, the Pervert, and Avant-Garde Opera of the *Fin de Siècle*," *New German Critique* 43 (Winter 1998), 35–68; Karen Painter, "The Sensuality of Timbre: Responses to Mahler and Modernity at the *fin de siècle*," *Nineteenth Century Music* 18.3 (Spring 1995), 236–56; Susan McClary, *Feminine Endings: Music, Gender and Sexuality* (Minneapolis: University of Minnesota Press, 1991); and a whole edition of the journal *Victorian Studies* in 1986.
3. Luigi Galvani, "De viribus electricitatis in motu musculari. Commentarius," *De Bonoiensi Scientiarum et Artium Instituo atque Academie Commentari* (1791), 7: 363–418. For an outline of the cultural reception of electricity see Christoph

Asendorf, *Batteries of Life: On the History of Things and Their Reception in Modernity*, trans. Don Reneau (Berkeley: University of California Press, 1993).

4. Sven Dierig, "Hirngespinste am Klavier: Über,chopinisierte' Nervensaiten im Berliner *fin de siècle*," in Michael Hagner, ed., *Ecce Cortex* (Göttingen: Wallstein, 1999), 26–50.
5. See picture in Christopher Forth, "Neurasthenia and Manhood in *fin-de-siècle* France," in *Cultures of Neurasthenia from Beard to the First World War*, ed. Marijke Gijswijt-Hofstra and Roy Porter (Amsterdam: Rodopi, 2001), 351.
6. Marcel Proust's father, Adrien Proust, gives a detailed account of how to electrocute patients' genitals in his book on neurasthenia. Adrien Proust, *Treatment of Neurasthenia*, trans. Peter Campbell Smith (London: Henry Kimpton, 1902), 198–9.
7. For a discussion of this with reference to nineteenth-century English novels, see Weliver, *Women Musicians in Victorian Fiction*.
8. "Zu den Ungeheuerlichkeiten unseres modernen Culturlebens gehört auch die Idee, dass jedes Kind aus besseren Ständen musikalische Bildung haben muss. Nach der Begabung wird nicht gefragt. Die Mode beansprucht es, dass die jungen Leute Clavier klimpern können . . . Bei mangelnden Talent ist das Clavierstudium eine inadequate Leistung, die, da sie die körperliche und geistige Kraft an und für sich schon erheblich in Anspruch nimmt, nicht selten Nervosität hervorrufen und wenn sie von talentlosen Mädchen zu intensiv oder widerwillig betrieben wird, den Grund zu schweren Nervenkrankungen legt." Richard Freiherr von Krafft-Ebing, *Über gesunde und kranke Nerven* (Tübingen: Verlag der Laupp'schen Buchhandlung), 59.
9. See for example, Plato, Book II of "The Republic," in Thomas Taylor, ed., *The Works of Plato*, 5 vols (Frome: The Prometheus Trust, 1996), 1: 305–39; Plato, Books II and III of "The Laws," in Taylor, ed., *The Works of Plato*, 2: 333–83; Suzanne Cusick, "Gendering Modern Music: Thoughts on the Monteverdi-Artusi Controversy," *Journal of the American Musicological Association* 46 (Spring 1993), 1–25; Eckhard John, *Musikbolschewismus: Die Politisierung der Musik in Deutschland 1918–38* (Stuttgart: Metzler Verlag, 1994); Jane Fulcher, "Melody and Morality: Rousseau's Influence on French Music Criticism," *International Review of the Aesthetics and Sociology of Music* 11.1 (June 1980), 45–57.
10. See Martin West, *Ancient Greek Music* (Oxford: Clarendon Press, 1992); Martin West, "Music Therapy in Antiquity," in Horden, ed., *Music as Medicine*, 51–68; Edward A. Lippman, "The Sources and Development of the Ethical View of Music in Ancient Greece," *The Musical Quarterly* 49.2 (April 1963), 188–209; Plato, *The Republic* and *The Laws*.
11. Grant Allen, *Physiological Aesthetics* (London: Henry King, 1877), 98.
12. Charles Baudelaire, *Richard Wagner in Paris* (Minden: Bruns, 1902), 77.
13. See Erwin Koppen, *Dekadenter Wagnerismus* (Berlin: Walter de Gruyter, 1973).
14. On philosophical and aesthetic aspects of late-nineteenth-century Decadence, and on Wagner's importance within the Decadent movement, see also Don LaCoss in this volume.
15. As Cheryce Kramer has shown, the more old-fashioned style of Mozart and Mendelssohn was played in asylums; the advanced harmonies of Wagner and his ilk were not regarded as suitable. Cheryce Kramer, "Soul Music as Exemplified in Nineteenth Century German Psychiatry," in *Musical Healing in Cultural Contexts*, ed. Penelope Gouk (Aldershot: Ashgate, 2000).

16. Alan Walker, *Franz Liszt*, 2 vols (Ithaca: Cornell University Press, 1987), 1: 152.
17. Qtd. in Asendorf, *Batteries of Life*, 156.
18. Horden, ed., *Music as Medicine*, 328.
19. Nicolas Slonimsky, *Lexicon of Musical Invective* (Seattle: University of Washington Press, 1965), 237.
20. Ernst Jentsch, *Musik und Nerven* (Wiesbaden: Bergmann, 1911), 119.
21. "Die Krankheit zeigt sich in Verkehrtheit der Neigungen, Perversität der Begierden und Wünsche und in dem vollständigen Mangel der sittlichen und socialen Gefühle." Jentsch, *Musik und Nerven*, 52.
22. "Eine hirnerschütternde Instrumentation und die schauerlichsten Dissonanzen, so dass, wie ein Musikkenner sagt, Gehörnerven, so dick wie die Schiffstaue dazu gehören." Theodor Puschmann, *Richard Wagner: eine psychiatrische Studie* (Berlin: Behr's, 1872), 30.
23. Thomas Mann, "Tristan," in *Stories of Three Decades*, trans. H.T. Lowe-Porter (London: Secker and Warburg, 1946), 141.
24. Mann, "Tristan," 151.
25. Mann, "Tristan," 152.
26. Mann, "Tristan," 152.
27. Gertrude Atherton, *Tower of Ivory*, 2 vols (Leipzig: Tauchnitz Verlag, 1910), 1: 202.
28. Atherton, *Tower of Ivory*, 2: 55.
29. Atherton, *Tower of Ivory*, 1: 205.
30. Atherton, *Tower of Ivory*, 2: 13.
31. Atherton, *Tower of Ivory*, 2: 103.
32. Atherton, *Tower of Ivory*, 2: 286.
33. George Moore, *Evelyn Innes* (New York: Appleton and Co., 1898), 172.
34. Moore, *Evelyn Innes*, 196.
35. Moore, *Evelyn Innes*, 325.
36. Moore, *Evelyn Innes*, 358.
37. See Hanns Fuchs, *Richard Wagner und die Homosexualität* (Berlin: Barsdorf, 1903) and Oskar Panizza, "Bayreuth und Homosexualität: eine Erwägung," *Die Gesellschaft: Monatsschrift für Literatur, Kunst und Socialpolitik* 11 (1895), 88–92.
38. "Ich finde, dass diese Musik unserm Wesen so sehr entspricht." Qtd. in Fuchs, *Richard Wagner und die Homosexualität*, 129.
39. Richard Freiherr von Krafft-Ebing, *Neue Forschungen auf dem Gebiete des Psychopathia Sexualis* (Stuttgart: Ferdinand Enke, 1890), 108.
40. Thomas Mann, "Tonio Kröger" in *Stories of Three Decades*, 105.
41. With the kind of hyperbolic denouement beloved by Mendes' generation, Friedrich castrates himself, burns down his fairy tale castle (a version of Ludwig's Wagnerian palace at Neuschwanstein) and finally has himself crucified at the Oberammergau Passion Play.
42. The letters include bills for fabric received; one bill is for 301 yards of satin in 13 colors, 190 yards of pink and blue cotton, and various articles of clothing and furnishings, totalling 3010 Florins. Daniel Spitzer, *Richard Wagners Briefe an eine Putzmacherin* (Vienna: Carl Konegen, 1906), 57–8.
43. "Für unsere Zeitalter jedoch, dem wohl die Leidenschaftlichkeit in der Beurteilung nacktester Menschlichkeit allgemach verloren gegangen ist, und das die Lehren eines Lombroso, Krafft-Ebing und Nordau der breitesten Allgemeinheit zugänglich macht; für unser Streben nach Erkenntnis jener

Beziehungen, welche die körperliche Natur des Menschen mit seinem geistigen Schöpfungen zu einem unlösbaren Ganzen verknüpfen; vor allem für den unbefangenen Biographen, der die Einheit von Psyche und Physis nicht geflissentlich übersehen darf, bedeuten diese sonderbaren Atlasbestellungen ein nicht zu unterschätzendes Beobachtungsmaterial." Spitzer, *Richard Wagners Briefe*, x–xi.

44. "Seichte Köpfe, Cyniker und schale Witzler hätten sich darüber lustig gemacht, als bekannt worden sei, dass Richard Wagner Atlashöschen, sowie Schlafröcke und Stiefelchen aus diesem Stoffe trage, und die Farbe je nach der Art seiner Kompositionen wechsle. Er aber behaupte, dass wenn Wagner nie einen Reim gestabt und nie eine Note geschrieben, sondern nichts Anderes geleistet hatte, als durch sein Beispiel die Wissenschaft auf den Einfluss des Atlas auf die Musikalische Gestaltungskraft aufmerksam zu machen, dies allein hingereicht haben würde, seinem Namen die Unsterblichkeit zu sichern. So viel sei gewiss, dass nunmehr der Atlas jedem Komponisten unentbehrlich geworden und ein noch verlässlicheres Reizmittel sei als Kaffee, Thee oder Tabak, je dass vielleicht wenn die gehörigen Versuche angestellt würden, eine Atlas-Narkose erzielt warden könnte." Daniel Spitzer, *Verliebte Wagnerianer* (Wien: Julius Klinkhardt, 1880), 120–1.
45. Joris-Karl Huysmans, *Against Nature*, trans. Robert Baldick (Harmondsworth: Penguin, 1959), 118.
46. George Gissing, *The Whirlpool* (London: Lawrence and Bullen, 1897), 28.
47. Gissing, *The Whirlpool*, 136.
48. Gissing, *The Whirlpool*, 202.
49. Gissing, *The Whirlpool*, 264.
50. Gissing, *The Whirlpool*, 315.
51. Thomas Mann, *Buddenbrooks*, trans. H.T. Lowe-Porter (New York: Alfred A. Knopf, 1924), 407–8.
52. Mann, *Buddenbrooks*, 412.
53. Mann, *Buddenbrooks*, 597.
54. Mann, *Buddenbrooks*, 531.
55. James Gibbons Huneker, *Overtures* (London: Ibister and Co., 1904), 93. Similarly, Huneker's 1902 story, "A Piper of Dreams," describes the composer Illowski (who has things in common with Wagner and Tchaikovsky; "disciples of Charcot explained his scores") as a "hypnotist, he conducted his orchestra through extraordinary and malevolent forests of tone." James Gibbons Huneker, "A Piper of Dreams," in *Melomaniacs* (New York: Charles Scribner's Sons, 1920), 41–5.
56. Leo Tolstoy, *The Kreutzer Sonata* (London: Walter Scott, 1898), 122.
57. For a discussion of Svengali more generally, see Daniel Pick, *Svengali's Web: The Alien Enchanter in Modern Culture* (New Haven: Yale University Press, 2000); Edgar Rosenberg, *From Shylock to Svengali* (Stanford: Stanford University Press, 1960).
58. George Du Maurier, *Trilby* (Oxford: Oxford University Press, 1995), 49.
59. Du Maurier, *Trilby*, 11.
60. Du Maurier, *Trilby*, 244.
61. See Sander Gilman, *Jewish Frontiers: Essays on Bodies, Histories and Identities* (New York: Palgrave Macmillan, 2003); Céline Kaiser and Marie-Luise Wünsche, eds, *Die Nervosität der Juden* (Paderborn: Ferdinand Schöningh, 2003).

62. Qtd. in Asendorf, *Batteries of Life*, 170–1.
63. Stephen McClatchie, *Analyzing Wagner's Operas: Alfred Lorenz and German Nationalist Ideology* (Rochester: University of Rochester Press, 1998); Derek B. Scott, "The Sexual Politics of Victorian Musical Aesthetics," *The Journal of the Royal Musical Association* 119.1 (1994), 95.

IV. The Traumatized Brain

7
Emergent Theories of Victorian Mind Shock: From War and Railway Accident to Nerves, Electricity and Emotion

Jill Matus

Historicizing trauma

In the last few decades, trauma theory has achieved great saliency in an array of disciplines. It has been widely applied in studies of twentieth-century forms of testimony and the capacity of literature to bear witness to traumatic experience, not just individual but generational and national. From the uniquely personal repercussions of childhood abuse to the wide-scale reverberations of colonial rupture, the concept of trauma has come to cover a wide range of suffering.[1] E. Ann Kaplan remarks that "it is partly because of accumulated twentieth-century traumatic events that psychologists, sociologists, and humanists are investigating trauma."[2] Although no one could claim that the twentieth century has the monopoly on horrific experience, trauma theory, it has been suggested, emerged as a response to "modernity." This view arises in large measure from the influential work of Walter Benjamin, which identified modernity with a rupture in experience and a break in consciousness.[3] But, as Benjamin himself understood, the material conditions and technologies we associate with modernity began well before the twentieth century.[4] Large-scale cataclysmic accidents, experiences of near death and miraculous survival were certainly part of the Victorian industrialized world. How did Victorians understand the effect on consciousness and memory of events and experiences that "went beyond the range of the normal"—events so overwhelming and inassimilable that the ordinary processes of registration and representation were suspended or superseded? And what proposed architecture of mind would support a theory of ruptured or suspended registration? At the same time that factory and railway accidents and war experience produced psychic effects demanding medical and legal attention, Victorians were also attentive to other kinds of shocks to the mind, which were less dependent on external cataclysm and more closely related to private, individual disruptions of consciousness and composure. Moreover, at least from the mid-century, Victorian physiologists exploring the nature of consciousness and the conundrum of nerve transmission came to figure the basic unit of consciousness itself as a kind of shock.

The term "traumata" to mean psychic wound was not used until 1894, when the *Oxford English Dictionary* (*OED*) notes its deployment in William James's summary of an early work of Sigmund Freud and Josef Breuer:

> 1894 W. James in *Psychol. Rev.* I. 199 Certain reminiscences of the shock fall into the subliminal consciousness, where they can only be discovered in 'hypnoid' states. If left there, they act as permanent 'psychic *traumata*', thorns in the spirit, so to speak.[5]

Histories of trauma generally focus on the advent of Freud, which is to say that the acknowledged roots of trauma theory lie in the psychoanalytic discourse of the early twentieth century—pre-eminently the work of Pierre Janet and Freud. Resisting the pronouncement that "it all begins with Freud," I want to ask what sort of earlier explanations of mind shock were formulated in the nineteenth century.[6] A recent trend in literary criticism deploying trauma theory is the retrospective diagnosis of anguished fictional subjects as suffering from trauma, as if trauma were a timeless and historically transcendent category. It is seldom that critics invoking the concept of trauma reflect on the history of its discursive development and provenance.[7] As an alternative to the application of contemporary or Freudian formulations of trauma (themselves vexed and continually under interpretative revision), my project is to explore what Victorian novelists and mental physiologists understood by psychic shock and what valence and social meanings that concept had in mid- to late-Victorian culture.

Such an exploration takes us from the large-scale and cataclysmic events of the external world to the smallest and most invisible transactions and transmissions of the internal world—from narratives of Victorian war and railway disaster to accounts of the physiology of feeling, sensation and emotion; to new ways of understanding the nerves and how they receive and transmit messages.[8] It is also a journey that I hope will allow us further purchase on how and why the dichotomy of active mind/passive body recurs so frequently in the cultural and medical history. As historians of psychology have noted, developing nineteenth-century scientific areas of inquiry (physiology, medicine) and so-called pseudoscientific areas (mesmerism, phrenology) propelled "the puzzle of the mind–body relation" to the forefront of public debate.[9] By attending to the literary text not only as an index of cultural reactions to scientific concepts, but also as an agent in developing discourses of the mind and body, we may at once broaden our sense of the complexity of Victorian formulations of emotion and their disruptive potential and provide further historicization of contemporary theories of emotion and of trauma.

War, railway accidents and cataclysm

Freud had been writing about the traumatic neurosis since the 1890s, but the case studies of shell-shocked soldiers in World War One prompted him to theorize the concept of trauma as overwhelming fright or shock. It is possible, therefore, that Victorian responses to war might yield a particularly Victorian discourse about the nature of overwhelming shock on the psyche. Both the Crimean War (1854–56) and the Boer War (1899–1902), in which numbers of British troops were involved, provide potentially useful starting points for an investigation of Victorian mind shock. Both involved large-scale suffering and casualties. Similarly, one might expect that the gruesome accounts of the so-called Indian Mutiny or Sepoy Rebellion of 1857, which horrified British subjects everywhere, would likewise document responses of shock to the slaughter and carnage experienced.

There has been some recent research by medical historians Edgar Jones and Simon Wessely on the occurrence of psychological disorders such as chronic fatigue syndrome in the Crimean War and Sepoy Rebellion. Examining the pension files of the Royal Hospital, Jones and Wessely focus on the case of one Charles Dawes, a soldier who served both in the Crimea and then in India in the suppression of the mutiny.[10] Dawes presented many symptoms that today would have qualified him for a diagnosis of chronic fatigue syndrome. The authors note too that "in the American civil war and the Boer war, soldiers exhibiting chronic fatigue syndromes, sometimes precipitated by the stress of battle, were occasionally diagnosed as suffering from the after effects of sunstroke."[11] They conclude that the "way we categorise symptoms and how we construct disorders is open to considerable variation, even if the clinical presentations are relatively stable."[12] In a subsequent article on psychiatric battle casualties they compare Victorian to later World War One and Two cases and conclude that Victorian incidences of psychosomatic disorder were either undiagnosed or somatized. "It appears, therefore, that unambiguous cases of combat fatigue were rarely identified in the

Victorian period and that soldiers traumatised by the stress of battle appear to have somatised their fears often in the form of disordered action of the heart (DAH) or psychogenic rheumatism."[13] With the identification of shell-shock and neurasthenia, the number of cases diagnosed as psychological in World War One rose a great deal. One reason that the authors of this study give for the increased focus on psychological issues is the reduced number of battle fatalities: "Advances in medical science have progressively reduced the proportion of troops who die from their injuries, the rate falling from 20% in the Crimean War to 6.1% in the First World War, 4.5% in the Second World War and 2.5% in Korea."[14] Improved survival rates may have "increased the number of potential psychiatric casualties and allowed the focus of attention to move towards psychological issues."[15] Another reason, I would suggest, reiterating that trauma and, for that matter, chronic fatigue syndrome or PTSD (Post Traumatic Stress Disorder) are historically produced categories rather than timeless pathological entities, is that the articulation of shell-shock and neurasthenia as "syndromes" may have provided a symptom language for sufferers, a form in which their disturbances could be expressed.

An obvious factor inhibiting the recognition of psychological disturbance as a result of war is the determined stoicism with which soldiers recounted their war experiences. The glory of war and the heroism associated with enduring pain of any kind made the confession of mental symptoms unlikely, and yet from time to time the memoirist allows us a glimpse of the horror and psychological toll exacted by war and, especially in the case of the Crimean War, the ravages of the cholera epidemic, which killed more soldiers than were claimed by battle injuries. In narratives of the Crimean War experience such as *Kars and our Captivity in Russia* (1856), the author writes movingly about the horrors of war as a "black and ghastly picture of horror and suffering that still clouds [his] . . . memory like the hovering phantoms of some hideous dream."[16] The prevailing interpretation in the narrative, however, is that exertion and the stress of war are stimulants. Cholera thus abates from "the moral exertion and physical excitement, and all the fierce energies stirred by the perils of the battle-field" but it creeps again upon the troops "when the roar of our cannon was hushed into the sad silence of expectation, and the dreadful despondency of hope deferred. . . . The sights and sounds by which we were surrounded were not such as to inspire men with the strength which may sometimes shake off a dangerous epidemic."[17] Of his General he notes that "as soon as the intense mental excitement of Kars was over, the wear and tear which his health had undergone, having no longer any sustaining stimulant, began to tell on him."[18] And as a prisoner of war, when the author does feel released from trying responsibilities and "risks or terrors to be endured" he nevertheless cannot sleep. So habituated has he become to a state of alarm, that now, in a good state of health, he is disturbed by every slight sound.[19]

Alexander Kinglake's *The Invasion of Crimea* (1863) places considerable emphasis on the fatigue, exhaustion and suffering of the troops during the winter months, but Kinglake shapes the narrative in terms of stoicism and heroism, the soldiers "bearing cold and hardships of all kinds with obstinate pride." He writes that "Without any extraneous aid men found strength, it would seem, in their own heroic qualities, found strength in that soldierly pride which forbids outward signs of disclosing self-pity or despair; and it is not, indeed, even certain that such of them as remained for the moment unstricken by illness were at all in a mournful humour."[20] "Misery ceases to be overwhelming when it is no longer solitary; individual loss is drowned in the feeling of common sympathy," writes James Gibson in his memoirs of the 1880s. "We are roused from the slumber of a lengthened peace to activity and energy in a noble cause; one feeling, and one alone, pervades all classes in the United Kingdom; one subject, and one alone, is the theme on which we delight to dwell—how we can best promote the safety of, and show our admiration for, the bravest army that ever left England's shores."[21]

Moving from memoirists to medical discourse, we find similar convictions about the rousing and even salubrious effect of emotion in times of war. Daniel Hack Tuke focuses on emotional response to war in his 1872 treatise on the effect of the mind on the body in health and disease. "Stirring political events, demanding individual action, have a wonderful influence over nervous affections. This fact was exhibited in the first American war." He cites authorities who state that "many whose habits were infirm and delicate, were restored to perfect health by the change of place or occupation to which the war exposed them."[22] Similarly, Tuke claims that "the battle-field constantly affords examples of the influence of an engrossing emotion in blunting sensation." He quotes from an article on the battle of Monte Rotundo (1867) in which a spectator wrote in the *Cornhill Magazine*: "All day long the battle raged; the troops were fainting with hunger and fatigue. Certainly they were the liveliest, most patient set of sufferers I ever saw; *the certainty of victory chloroformed their pain*" (original emphasis).[23]

A different perspective on emotional, nervous or apparently non-physical injury is offered in *The Crimean Journals of the Sisters of Mercy, 1854–56*, which allows a fascinating glimpse into the tensions between Florence Nightingale and the Catholic Sisters with whom she had to work at the front. The various accounts of nursing rivalries include a contrast between the hardy, capable sisters, inured to shock, and their more delicate, less efficient secular counterparts, as well as a sense of psychological or spiritual injury in the troops. Sister M. Aloysius Doyle notes in a journal from Scutari that "the secular ladies" are unable to withstand the shocks of war which the Sisters can. "No wonder Miss. Nightingale should have leaned on Mother Mary Clare, of Bermondsey, and her four Sisters."[24] Highly critical of Nightingale, Mother Francis Bridgeman implies in her journal that Miss. Nightingale had no sense of injury beyond the physical: "Now at this time there were about four thousand patients in Scutari hospitals, from fifty to ninety were daily buried. We found the sufferers of *that year peculiarly prostrate in mind and body*; yet Miss. Nightingale coolly affirmed they needed no nursing, as they were not wounded!" (Original emphasis.)[25] In Mowbray Thomson's *The Story of Cawnpore*, some of the more frail women involved in the conflict were reputed to have died of shock, but such statements speak more to gendered notions of emotional susceptibility than they imply any recognition of specifically psychic or emotional injury.[26]

If British sources on the Crimean War and the Sepoy Rebellion turn up relatively little overt discussion of psychological or emotional injury, there is somewhat more to be found on this subject in the medical discourse of the 1870s surrounding the siege of Paris during the Franco-Prussian War. On reflection, the reason for this becomes obvious. Since the valor and heroism of British troops are not at stake, observers are unconstrained in speculating about French susceptibility to emotional disturbance. Indeed, one might argue that national stereotypes of the French character as unstable and excitable (or worse) may even play into the numerous comments on the rise of psychological problems following the siege of Paris. The author of an article on "The War" in January 1871 in *The Lancet* noted that "the number of those who have become insane since the war is very great; the asylums are quite full."[27] Later that year, it was observed that after the city's capitulation "the mental shock to some was such that they almost lost their reason!"[28] In the *Journal of Mental Science* (1872), an anonymous article entitled "Effects of Fright on the Mind" makes reference to the "late siege of Paris" as furnishing some interesting examples of the effects of a "profound shock on the mind."[29] The author quotes a French medical source, which reported: "On July 14th last, about 1.45, a tremendous explosion resounded throughout Paris. The percussion cap manufactory at Vincennes exploded. . . . The next day a very intelligent lady, who had witnessed the whole affair, was found to have *no recollection whatever of the occurrence*. A great emotional disturbance may then, in a state of health, be effaced from the memory" (original emphasis).[30] Looking back in 1892 at the history of "shock from fright," the railway physician Herbert Page, who was one of the first to draw attention to the psychical disturbance that terror in railway accidents could produce, noted that

> [t]he siege of Strasbourg and the siege of Paris during the last Franco-German war [1870–71] were both productive of many examples of *grave nervous disorder*, even ending fatally, which clearly had their origin in the terrible circumstances to which the sufferers had been exposed—to wit, the constant bursting of shells, the ever-present sense of danger, the anxiety as to the safety of friends, the inadequacy of the food supply. Happily in this country we have been spared such experiences, but like *sources of neurotic disturbance* are to be found very often in the events of an ordinary railway collision, where we have in combination everything which is likely to induce great terror—magnitude and violence of the forces, loud noise, shrieks of the injured and utter helplessness of individual passengers. (Emphasis added.)[31]

Page's comparison of the French responses to war and the reactions "in this country" to railway collisions draws attention to the now well-documented discourse surrounding railway accidents from the 1860s to the end of the century. Railway accidents, even more than war, brought strange cases to medical attention and provoked a range of diagnoses (such as railway spine and railway brain). Indeed, it has not escaped the notice of historians of railway accidents that the medical discourse of shock was given a jolt by the railway disasters of the 1850s and 1860s and the controversial injuries (and insurance claims) to which they gave rise. One of the earliest specific treatises on the subject, Edwin Morris's *A Practical Treatise on Shock* (1867) was written in response to the phenomenon of railway shock. The full title of Morris's treatise is *A Practical treatise on Shock after surgical operations and injuries; with especial reference to shock caused by railway accidents*. And in his preface, Morris warns that his chapter on the railways is included to "assist in unraveling those intricate cases in which there is every reason to believe the symptoms are simulated, and at the same time to put medical men on their guard against such cases."[32]

Though overtly suspicious of malingerers, Morris does, at the outset, place emphasis on both mental and physical forms of shock, and begins by describing shock as "that peculiar effect on the animal system, produced by violent injuries from any cause, or from violent mental emotions—such as grief, fear, horror or disgust."[33] Though he pays most attention to gunshot wounds and injuries both proximate to and distant from the brain, he returns in his conclusion to the way in which the brain and nervous system are paralyzed by shock, and "volition and sensation are temporarily suspended."[34] Morris may be hedging his bets with his inclusive sweep of all violent physical injury or emotional experience as possible causes of shock; yet his attention to emotional causes is part of a continuous thread in nineteenth-century scientific and pseudoscientific discourse giving at least cautious credibility to the power of emotions.

In a recent study of the law relating to liability for negligence resulting in psychiatric injury, Danuta Mendelson observes that while medical discourse has evolved to produce an array of terms and concepts to express changing ideas of psychical injury, law has kept to the basic, one might say generic, term, "nervous shock," which is still used in British, Australian and Canadian law. This adherence produces some confusion because "nervous shock" is no longer a diagnostic category in contemporary medicine and psychiatry and does not have the medical meanings that it once carried. While, as Mendelson notes, this produces frustration in lawyers, it is instructive for historians of the psychic wound because it suggests that late-nineteenth-century jurisprudence, beginning to take account of the need to compensate sufferers for psychic harm, was looking for a terminology that indicated psychological injury but also implied a physical or bodily anchor. According to Mendelson, the first case in which a plaintiff was awarded damages for nervous shock involved the witnessing of a railway accident—*Coultas v. Victorian Railway Commissioners* (1886).[35] The importance of railway disaster and consequent insurance claims and legal pronouncement to the development of conceptions of psychic injury cannot be underrated and has justly received critical attention.[36] What has been rather less explored, but is also significant is an earlier, arguably preparatory, discourse on the nature of consciousness, which conceptualizes shock at the micro-level of nerves, consciousness and emotion.

Nerves and shock

An example from George Eliot's novel *Middlemarch* (1872) will serve to introduce some key issues in this discourse. Dorothea Brooke is on her unhappy wedding journey in Rome, where "underwhelmed" by the rector Edward Casaubon, her new husband, she is nevertheless overwhelmed by the carnival of sights and sounds that Rome represents. The "stupendous fragmentariness" of Rome is described as heightening the "dream-like strangeness" of her bridal life.[37] Rome, the "city of visible history" is an experience whose effect on consciousness and memory George Eliot likens to that of an electric shock:

> Ruins and basilicas, palaces and colossi, set in the midst of a sordid present: . . . all this vast wreck of ambitious ideals, sensuous and spiritual, mixed confusedly with the signs of breathing forgetfulness and degradation, at first *jarred her as with an electric shock,* and *then urged themselves on her with that ache belonging to a glut of confused ideas which check the flow of emotion.* Forms both pale and glowing *took possession* of her young *sense, and fixed themselves in her memory even when she was not thinking of them,* preparing strange associations which remained through her after-years. Our moods are apt to bring with them images which succeed each other *like the magic-lantern pictures* of a doze. (Emphasis added).[38]

There are several elements in the description relevant to the topic of mental shock: the conceptualizing of strong emotion as a jolt of electricity, which gestures toward the mid-nineteenth-century scientific context in which nerve force was increasingly understood in electrical terms; the ensuing imprint on memory and possession of the senses; the return of the pictures beyond conscious recall—a magical technology of the unconscious that produces a ghostly and marvelous parade of images succeeding each other in states of slightly altered consciousness, light sleep, doze or dream. George Eliot also refers to Dorothea's state as one of "inward amazement," which, coupled with the earlier description of the "dream-like strangeness of her bridal life" suggests that this profound experience is tantamount to an altered state of consciousness like dream or trance. In focusing on the shock of this experience, I am not suggesting that George Eliot presents Dorothea's experience here as a disabling wound. There are ways in which this experience is a necessary shaking up of her heroine's narrow, provincial world. Nevertheless, it is a jarring experience that possesses her and controls her memory involuntarily—haunts her, we might say.

George Eliot's evocative account simultaneously draws on and helps to construct a particularly Victorian discourse of the way in which mind and body are affected by powerful emotional experience. Such discourse is the seed-bed (to invoke an Eliot-like metaphor) for the formulations of psychic trauma that emerge at the end of the nineteenth century.

Literature is important to the historicization of ideas about trauma precisely because trauma is a culturally and historically constructed concept; the medical and psychological discourse that produced it ought to be seen in relation to its ambient culture.[39] As Janet Oppenheim puts it in her history of the discourse of nerves, "scientists and medical doctors, belonging integrally to the public . . . , share many of its biases and expectations. Their pronouncements are not objective, or free of implicit moral judgment, for science and medicine are interpretative endeavors into which the surrounding social context constantly intrudes."[40] I would add "cultural" to "social" in drawing attention to literature's status as a complex cultural document and an influential shaper of public opinions and perceptions.

George Eliot's attention to the jolt Dorothea's consciousness receives from a powerful experience is consistent with the way physiological psychologists such as Herbert Spencer and Alexander Bain were thinking of consciousness itself. Indeed, both offer definitions of consciousness as a form of shock. In his *Principles of Psychology* (1855) Spencer argues that it is possible, "may we not even say probable—that something of the same order as that which we call a nervous shock is the ultimate unit of consciousness; and that all the unlikenesses among our feelings result from unlike modes of integration of this ultimate unit."[41] Such shocks, he goes on to explain, are faint pulses rather than powerful charges; if we felt each one of them consciously, we would not be able to cope:

> Were our various sensations and emotions composed of rapidly-recurring shocks as strong as those ordinarily called shocks, they would be unbearable: indeed life would cease at once. We must think of them rather as successive faint pulses of subjective change, each having the same quality as the strong pulse of subjective change distinguished as a nervous shock.

Spencer concludes that if "the subjective effect or feeling, is composed of rapidly-recurring mental shocks … it corresponds with the objective cause—the rapidly recurring shocks of molecular change."[42] If sensation itself is conceptualized as a tiny unregistered shock, then it follows that a very powerful feeling would register as a palpable one. This point is explicitly made by Alexander Bain in *The Emotions and the Will* (1859). He writes that a "sudden shock of feeling is accompanied with movements of the body generally, and by other effects. When no emotion is present, we are quiescent; a slight feeling is accompanied with slight manifestations; a more intense shock has a more intense outburst. Every pleasure and every pain, and every mode of emotion, has a definite wave of effects, which our observation makes known to us."[43] Bain's point is that emotion is itself a kind of shock that produces waves which transmit effects throughout the body in proportion to the intensity of the initial shock.[44] Reiterating this position in his later work *Mind and Body* (1873), Bain writes:

> The simplest term we can employ for a mental state is a shock; a word equally applicable to the bodily side and to the mental side. A sudden stimulation of the eye, the ear, the skin, the nose, is called a shock, from its more outward or physical aspect; it is also called a shock mentally . . . because there is a rapid transition from quiescence to excitement, in which circumstance there is an acute parallelism between the otherwise distinct physical and mental facts.[45]

The concept of shock articulated here is the great solvent of the divide between body and mind. Around the same time, Daniel Hack Tuke, who, as we have seen, focused on the relations of body and mind in the conception of disease and treatment, declared that

> an emotion may also be conceived to cause structural change in the higher centres of the encephalon. . . . It is easy to see how, from Fright or sudden Joy, there may be a shock, more or less temporary, to the motor centres, by which some part is rendered unable to respond to the stimulus of the Will, or of ideas, or emotions, just a man is sometimes deaf for days after firing a cannon, or is blind for a time after his eyes have been subjected to intense light.[46]

He concludes that "[p]robably all we can say with certainty is that the shock which the brain receives from a violent emotion like Terror disturbs the normal relative nutrition and vascularity of the volitional and motorial centres."[47] Common to all these examples is the underlying assumption that what we experience as mental effect always has a physiological correlative.

Shock in the texts I have been quoting is likened to tremor and earthquake, a violent disruption, a clash, or physical upheaval. Mechanical and climatic, it can also be imaged in terms of electricity, as we saw in George Eliot's description of Dorothea. During the last half of the nineteenth century, especially in Germany, electricity was being established as the agent of nerve force. Up to date on recent discoveries in this regard, George Eliot writes in *Daniel Deronda* (1876) of the average man, unaware of many important facts about himself, and dark to those "even concerning the action of his own heart and the structure of his own retina. A century ago he and all his forefathers had not had the slightest notion of that electric discharge by means of which they had all wagged their tongues mistakenly. . . ."[48] According to psychologist Keith Oatley, the significant discovery in brain research at the end of the eighteenth century was that nerves worked not by hydraulics but by electricity.[49] But it is important to remember that electricity in this regard is still being thought of in the context of a fluid. By the end of the eighteenth century, the idea of "an electrical fluid flowing in nerve tubules was an attractive hypothesis as the agent of nerve action," though, as Sidney Ochs points out, it was far from universally accepted.[50] Still, the analogical force of the battery was powerful in physiological texts such as William Carpenter's *Principles of Mental Physiology* (1874):

> Just as a perfectly constructed Galvanic battery is *inactive* while the circuit is "interrupted," but becomes *active* the instant that the circuit is "closed," so does a Sensation, an Instinctive tendency, an Emotion, an Idea, or a Volition, which attains an intensity adequate to "close" the circuit, liberate the Nerve-force with which a certain part of the Brain, while in a state of wakeful activity, is always charged. (Original emphasis.)[51]

The history of how electricity came to be accepted as the agent of nerve action is a long one, extending from the ancients, who noticed that when amber (the Greek name for electricity) was rubbed it would attract small objects to it, to the work of Galvani at the turn of the eighteenth century, "when electricity was generated as a static discharge and its potent effects on the body experienced."[52] Galvani's explanation was that the nerve fibers were so constituted that they were hollow internally or composed of material able to carry electric fluid, but were oily externally so that the electric fluid was prevented from dissipating. Galvani posited that electric fluid was produced in the cerebrum and extracted from the blood, entering the nerves and circulating within them. He suggested that this animal electricity was identical to animal spirits, and hoped that if this were the case, there would at last be clarity on that opaque subject.

German physiologist Emil du Bois-Reymond, known as the father of experimental electrophysiology, posited that current flowed continually in the fibers of muscle and nerves.[53] Though this proposition was erroneous, as later research into the electrical nature of nerve and muscle potentials in the 1880s showed, du Bois-Reymond was on the right track and also made important strides in measuring the conduct velocity of nerve impulse. In *Mind and Body*, Alexander Bain refers repeatedly to "nerve-electricity" and makes mention of du Bois-Reymond's computation of nerve impulse transmission from the location of a harpoon wound to the brain in a whale; he also refers to Hermann Helmholtz's attempts to calculate reflex action in frogs.[54] Even earlier, as the following example from Morris's medical treatise on shock (1867) illustrates, general acceptance of the nature of electricity as an agent of nerve action provided a way of understanding shock:

> Of the active principle in the nervous system, we in reality know little or nothing; this, however, is a matter of no importance to us, and I only allude to it, in consequence of the important part it plays in the production of the train of symptoms called *shock*. I am inclined to think that it is of the same nature as electricity, and the nerves act as conductors. Faraday has propounded this idea, and Abernathy also advocated this doctrine; certain it is that this nervous agent has an action of a peculiar kind, and if not identical with electricity, is very analogous to it.[55]

Only toward the end of the nineteenth century was the electrical nature of nerve potential better grasped when the properties of ions in electrolyte solutions were demonstrated by the Swedish physical chemist, Svante Arrhenius.[56]

The physiology of emotion

If electricity was the means by which nerves transmitted impulses, then was it possible literally to experience a powerful transmission of nervous energy as an electric shock? While George Eliot could effectively represent shocks to consciousness as electric, the question of what actually goes on inside the body when emotions are experienced was one that greatly occupied physiological psychology at mid-century. Indeed, the mid-nineteenth century is often seen in histories of emotion theory as *the* moment when emotion was conceived of as a bodily, physiological process. Historians of psychology, such as Kurt Danziger, have argued that the increasing physicalist tendency in psychology and physiology was behind the discursive formation of the term "emotion" which became standard around the mid-century.[57]

More recently, Thomas Dixon, in *From Passions to Emotions: The Creation of a Secular Psychological Category*, has claimed: "[t]he category of emotions conceived as a set of morally disengaged, bodily, non-cognitive and involuntary feelings, is a recent invention." This statement does not, however, do justice either to the debate surrounding the nature of emotion in the nineteenth century or the vigorous and ongoing twenty-first century forms of that debate. Dixon suggests that "prior to the creation of the emotions as an over-arching category, more subtlety had been possible on these questions. The 'affections' and the 'moral sentiments,' for example, could be understood as both rational and voluntary movements of the soul, while still being subjectively warm and lively psychological states."[58]

The central question of Dixon's study is why during the first half of the nineteenth century a "wholesale change in established vocabulary occurred such that those engaged in theoretical discussions about phenomena including hope, fear, love, hate, joy, sorrow, anger and the like no longer primarily discussed the passions or affections of the soul, nor the sentiments, but almost invariably referred to 'the emotions.'"[59] Just as Ian Hacking has argued in *Rewriting the Soul: Multiple Personality and the Sciences of Memory* that memory science of the last quarter of the nineteenth century was a means of scientizing the soul, so Dixon argues that the substitution of "emotion" for older terms reveals a similar scientizing of the soul—a secularization of psychology. He argues that really all of Bain, Spencer and Darwin (and one could certainly include George Henry Lewes in this company) inclined toward an epiphenomenalist view of mind—that is "the real business of emotions went on at the physiological and neurological levels." Emotions appeared to be the mental side of what was really and objectively an activity of the central nervous system.[60] Thus Alexander Bain in *Mind and Body*: "When a shock of fear paralyses the digestion, it is not the emotion of fear, in the abstract, as a pure mental existence, that does the harm; it is the emotion in company with a peculiarly excited condition of the brain and nervous system; and it is this condition of the brain that deranges the stomach."[61]

According to Dixon's explanation of the Victorian physiology of emotions, the mind becomes increasingly passive as greater agency is attributed to the body.[62] The terms in which he characterizes the history of emotion hold in place the old dualism by which the body or viscera become primary and the "brain and its mind," as he puts it, become secondary.[63] Time and again, however, Bain talks of the interrelationship between mind and body, not the reduction of the former to the latter.

Although he does aver that "the mind is completely at the mercy of the bodily condition" he does so in order to show that "there is no trace of a separate, independent, self-supporting, spiritual agent, rising above all the fluctuations of the corporeal frame."[64] Whatever mind is, he implies, it is thoroughly tied up with body. In this way, Bain attempts to dismantle the dualism of mind and body, but not to reduce mind to brain, or to suggest that mind is merely and passively body. Indeed, many nineteenth-century physiologists worked assiduously to adjudicate that dualism as non-competitive. Contesting the derisive label "materialist," physiologists such as William Carpenter opted for a position known as dual-aspect monism, in which bodily processes were one side of the same coin as feelings and subjective experiences of those processes and states. And of George Henry Lewes, it has been noted that "*The Physical Basis of Mind* (1877) . . . elaborates [his] . . . important theory of the relationship between—and in a specific and fundamental sense even the identity of—mind and body. Lewes contends that physiology and psychology, nerves and neuroses, are best understood as, respectively, the objective and subjective presentations of what are in fact the same phenomena . . . What *looks* like physiological excitation to the objective observer may *feel* like thought or sensation to the subject, but neither point of view is secondary or reducible to the other."[65]

It is worth emphasizing that a commitment to the physiology of emotion is not necessarily an abandonment of the cognitive, evaluative capacity of emotion. Drawing once again on the vocabulary of electric shock, George Eliot in *Middlemarch* illustrates both the physiological and cognitive aspects of strong emotional experience in the following description of the effect of Dorothea on Casaubon's young cousin, Will Ladislaw:

> When Mrs. Casaubon was announced he started up as from an electric shock, and felt a tingling at his finger-ends. Any one observing him would have seen a change in his complexion, in the adjustment of his facial muscles, in the vividness of his glance, which might have made them imagine that every molecule in his body had passed the message of a magic touch. And so it had. For effective magic is transcendent nature; and who shall *measure the subtlety of those touches which convey the quality of soul as well as body*, and make a man's passion for one woman differ from his passion for another as joy in the morning light over valley and river and white mountain-top differs from joy among Chinese lanterns and glass panels? Will, too, was made of very impressible stuff.[66] (Emphasis added.)

The passage is striking as evidence of the way emotion, internal feeling, is rendered bodily and physical, read through its corporeal manifestations. In this regard, George Eliot is very much in line with the way emotions themselves in physiological psychology were being emphasized as a physical process. But in conjunction with the bodily changes Will experiences, we are encouraged to recognize the magical aspect of the transformation—even as George Eliot details the physical process, she keeps alive a sense of the miraculous, marvelous and inexplicable. And she preserves the language of soul. This is not, however, a remystification. If Will's emotional response is visceral it is also cognitive—the passage speaks of his evaluation and assessment of Dorothea's worth; his discrimination of her value among other women. To express the small but significant choices, preferences and discriminations that reveal our cognitive schemes, George Eliot contrasts a preference for natural light and natural scenery with artful light and manufactured materials. As in the passage I quoted earlier from *Middlemarch*, the lantern (here Chinese, there magic) is an important metaphor allowing George Eliot to draw on visual technologies to represent internal processes such as cognitive choice and involuntary memory. Finally, the statement about Will being made of impressible stuff—a good conductor of the emotional as electrical—is a light, slightly ironic, touch, bringing us (and him) down to earth after the sublimity and seriousness of the previous remarks on body and soul.

Martha N. Nussbaum has recently argued in *Upheavals of Thought: The Intelligence of Emotions* that emotions have come to be regarded as "'bodily' rather than 'mental,' as if this were sufficient to make them unintelligent rather than intelligent." She continues: "Although I believe that emotions are, like other mental processes, bodily, I also believe, and shall argue, that seeing them as in every case taking place in a living body does not give us reason to reduce their intentional/cognitive components to nonintentional bodily movements. . . . Certainly we are not left with a choice between regarding emotions as ghostly spiritual energies and taking them to be obtuse, non-assessing bodily movements, such as a leap of the heart, or the boiling of the blood."[67] Anticipating Nussbaum's refusal of binary oppositions, George Eliot does not structure her representation of the power of emotion in terms of a dichotomy between physical and mental, bodily and cognitive, but attempts a kind of synthesis. Read in relation to the history of Victorian psychological discourse and the debates that continue in contemporary emotion theory, literary texts such as George Eliot's *Middlemarch* offer a more complex representation of Victorian responses to the nature of emotions than has generally been allowed.

Conclusion

The focus in this chapter on emotions and consciousness as nineteenth-century antecedents of a discourse of trauma may at first glance seem out of step with the preoccupations of much contemporary trauma theory, which places a great deal of emphasis on dissociated and dysfunctional memory. Although there is a growing body of recent work in the neurosciences exploring the neurophysiological mechanisms responsible for emotional consciousness and mapping the effects of stress on the hippocampus and amygdala, this work has not been integrated into humanist and social scientific approaches to trauma.[68] Neuropyschoanalyst Allan Schore has described the most far reaching effect of trauma as the loss of ability to regulate feelings, yet affect regulation has been eclipsed by memory in dominant and influential trauma theory over the past few decades.[69] Theorists, analysts and therapists such as Cathy Caruth, Dori Laub, Shoshana Felman and Judith Herman all emphasize unprocessed memory as the key element of trauma. And in his history of multiple personality disorder, *Rewriting the Soul*, which is, along the way, also an account of how trauma was constructed as a syndrome, Ian Hacking traces the rise of memory science in the last quarter of the nineteenth century. On the literary front, Nicholas Dames's remarks about trauma in *Amnesiac Selves* assume that its *sine qua non* is memory disturbance. He has argued that the Victorian novel (certainly before 1870) evinces an "amnesiac self" in the sense that memories deemed unproductive are expunged or absent from the narrative. Only after memory becomes nervous, a result of the burgeoning physiological psychology in the latter half of the century, do we find evidence of what Benjamin has called the "traumatophile."[70] The emphasis on memory as somehow the key to trauma arises understandably out of the puzzling question: how are cognitive processes affected in the event of overwhelming experience? Just as that question is answered in terms of unconscious rather than conscious registration, and involuntary rather than voluntary recall, so too the domain of trauma becomes the domain of memory. But in nineteenth-century discourse, disturbance of memory as the result of overwhelming or shocking experience is only one aspect of a network of related ideas that later cohere (albeit uneasily) as the concept of trauma.[71] In exploring that "seed-bed" of discourses, I would argue that it is important not to look singularly at memory, but more generally, as I have attempted here, at theories of consciousness, nerve force and emotion.

Notes

1. See the introduction to Ruth Leys, *Trauma: A Genealogy* (Chicago: University of Chicago Press, 2000).
2. E. Ann Kaplan, *Trauma Culture: The Politics of Terror and Loss in Media and Literature* (New Brunswick, New Jersey: Rutgers University Press, 2005), 24.
3. Walter Benjamin influentially drew attention to the "traumatophile" in his essay "On some motifs in Baudelaire." See Walter Benjamin, *Illuminations*, trans Henry Zohn, ed. and intro. Hannah Arendt (New York: Schoken Books, 1988), 155–200.
4. Benjamin drew particular attention to the way Baudelaire and Poe register the shock of city life; there are many British writers of the mid- to late-nineteenth century (for example, Dickens, George Eliot, Elizabeth Gaskell) attentive in different ways to similar accumulating pressures.
5. Ian Hacking, *Rewriting the Soul: Multiple Personality and the Sciences of Memory* (Princeton: Princeton University Press, 1995), 183 notes that the term "moral trauma"—*traumatisme moral*—was "already in circulation in 1885 . . . when Freud arrived in Paris to study under Charcot."
6. Kaplan, *Trauma Culture*, 25. She does add, however, that "along with Freud's contemporaries and the pioneering clinicians who preceded him nineteenth- and early twentieth-century thinkers had discovered the basic phenomena and structures of trauma."
7. I discuss this tendency in "Historicizing Trauma: The Genealogy of Psychic Shock in *Daniel Deronda*," forthcoming in *Victorian Literature and Culture*.
8. See Wolfgang Schivelbusch, *The Railway Journey: The Industrialization of Time and Space in the Nineteenth Century* (Berkeley: University of California Press, 1977) for an account of the medical treatment of shock in relation to railway accidents.
9. See Roger Smith, "The Physiology of the Will: Mind, Body, and Psychology in the Periodical Literature, 1855–1875" in *Science Serialized: Representations of the Sciences in Nineteenth-Century Periodicals*, ed. Geoffrey Cantor and Sally Shuttleworth (Cambridge, Massachusetts: The MIT Press, 2004), 81–110, 99.
10. Edgar Jones and Simon Wessely, "Case of Chronic Fatigue Syndrome after Crimean War and Indian Mutiny" in *British Medical Journal* 319 (1999), 1645–7. The authors have recently published *Shell-Shock to PTSD: Military Psychiatry from 1900 to the Gulf War* (Hove: Psychology Press, 2006).
11. Jones and Wessely, "Case of Chronic Fatigue Syndrome," 1647.
12. Jones and Wessely, "Case of Chronic Fatigue Syndrome," 1647.
13. Edgar Jones and Simon Wessely, "Psychiatric Battle Casualties: An Intra- and Interwar Comparison" in *The British Journal of Psychiatry* 178 (2001), 242–7, 242.
14. The authors here (246) cite G.W. Beebe and M.E. DeBakey, *Battle Casualties: Incidence, Mortality, and Logistic Considerations* (Springfield: Charles C. Thomas, 1952), 77.
15. Jones and Wessely, "Psychiatric Battle Casualties," 246. See also Mark Micale's essay on the American context of psychiatric injury in this volume.
16. Colonel Atwell Lake, C.B., *Kars and Our Captivity in Russia* (London: Richard Bentley, 1856), 27.
17. Atwell Lake, *Kars and Our Captivity in Russia*, 26.
18. Atwell Lake, *Kars and Our Captivity in Russia*, 263.
19. Atwell Lake, *Kars and Our Captivity in Russia*, 245–6.

20. Alexander Kinglake, *The Invasion of Crimea: Its Origin, and an Account of its Progress Down to the Death of Lord Raglan* (1863), vol. 4, *The Winter Troubles*, 6 vols. (New York: Harper & Brothers, 1881), 152.
21. James Gibson, *Memoirs of the Brave: A Brief Account of the Battles of the Alma, Balaklava, and Inkerman, with Biographies of the Killed and A List of the Wounded* (1889; London: The London Stamp Exchange, 1890), 27.
22. Daniel Hack Tuke, *Illustrations of the Influence of the Mind upon the Body in Health and Disease: Designed to Elucidate the Action of the Imagination* (1872; Philadelphia: Henry C. Lea's Son & Co., 2nd edn, 1884), 386.
23. Tuke, *Illustrations of the Influence of the Mind upon the Body*, 164.
24. *The Crimean Journals of the Sisters of Mercy, 1854–56*, ed. Maria Luddy (Dublin: Four Courts Press, 2004), 28.
25. *The Crimean Journals of the Sisters of Mercy*, 135.
26. Andrew Ward, *Our Bones are Scattered: The Cawnpore Massacre and the Indian Mutiny of 1857* (New York: Henry Holt, 1996); Ward quotes from Mowbray Thomson in noting that "many of the most frail and frantic of the women and children had died of shock and exhaustion" (Mowbray Thomson, *The Story of Cawnpore* [London, 1859], 66; qtd. in Ward, 203).
27. "The War" in *The Lancet* (21 January 1871), 98–9, 99.
28. Alex Boggs, "Paris after the Double Siege," in *The Lancet* (July 8, 1871), 75.
29. Anonymous, "Effects of Fright on the Mind," *Journal of Mental Science* 18 (July 1872), 234–5.
30. "Effects of Fright on the Mind," 235.
31. Herbert Page, "Shock from Fright" in Daniel Hack Tuke, ed., *A Dictionary of Psychological Medicine* (Philadelphia: P. Blackiston, 1892), 1157–60, 1157.
32. Edwin Morris, *A Practical Treatise on Shock* (1867; Philadelphia: J.B. Lippincott, 1868), v–vi. Contrast Page's view some 25 years later that "the early and amicable settlement of the claim for compensation on account of the injuries sustained" is one of the best treatments for disorders following railway shock; see "Shock from Fright," 1160.
33. Morris, *A Practical Treatise on Shock*, 9.
34. Morris, *A Practical Treatise on Shock*, 88.
35. See Danuta Mendelson, *The Interfaces of Medicine and Law: The History of the Liability for Negligently Caused Psychiatric Injury (Nervous Shock)* (Aldershot: Ashgate, 1998), 1–5. See also her discussion of *Coultas et uxor v. The Victorian Railway Commissioners*, 60–70.
36. This topic has received attention in histories of accident and trauma; see Ralph Harrington, "The Railway Accident: Trains, Trauma and Technological Crisis in Nineteenth-Century Britain" (1999), http://www.york.ac.uk/inst/irs/irshome/papers/rlyacc.htm (accessed January 21, 2007); see also Herbert Page, *Injuries of the Spine and Spinal Cord* (London: Churchill, 1883); and *Railway Injuries, With Special Reference to Those of the Back and Nervous System, in Their Medico-Legal and Clinical Aspects* (London: Griffin, 1891); Ruth Leys, *Trauma: A Genealogy*; Ian Hacking, *Rewriting the Soul: Multiple Personality and the Sciences of Memory* (Princeton: Princeton University Press, 1995).
37. George Eliot, *Middlemarch*, ed. David Carroll (1871–2; Oxford: World's Classics, 1998), 20: 190.
38. George Eliot, *Middlemarch*, 20: 191.
39. On the historicity and constructedness of trauma as a concept, see particularly Allan Young, *The Harmony of Illusions: Inventing Post-Traumatic Stress Disorder*

(Princeton: Princeton University Press, 1995); Ian Hacking, *Rewriting the Soul*; and Ruth Leys, *Trauma: A Genealogy*.

40. Janet Oppenheim points to the way scientists and doctors are "moulded, too, by systems of values, ethical codes, religious beliefs, and all manner of preconceived opinion." See *"Shattered Nerves": Doctors, Patients and Depression in Victorian England* (New York: Oxford University Press, 1991), 4.
41. Herbert Spencer, *The Principles of Psychology* (1855), 2nd edn, 2 vols (London: Williams and Norgate, 1870–2), 151. Although Michael Maher asserts that Spencer sees "all forms of consciousness . . . [as] resolvable into elementary units of feeling akin to electric-shocks," Spencer does not specifically refer to the shock as electric. See Maher's anti-materialist *Psychology Empirical and Rational*, 8th edn (London: Longmans, Green, 1915), http://www2.nd.edu/Departments///Maritain/etext/psych023.htm (accessed January 21, 2007).
42. Spencer, *Principles of Psychology*, 152.
43. Alexander Bain, *The Emotions and the Will* (1859), 3rd edn (London: Longmans, 1880), 4.
44. Bain, *The Emotions and the Will*, 10.
45. Alexander Bain, *Mind and Body*: *The Theories of their Relation* (London: Henry King, 1873), 40.
46. Tuke, *Illustrations of the Influence of Mind upon Body*, 246.
47. Tuke, *Illustrations of the Influence of Mind upon Body*, 230.
48. George Eliot, *Daniel Deronda*, ed. Graham Handley (1876; Oxford: World's Classics, 1984), 37: 402. George Henry Lewes was actively involved in experiments with frogs and electricity as he explored the physiology of consciousness; see Richard Menke, "Fiction as Vivisection: G.H. Lewes and George Eliot," *ELH* 67.2 (2000), 617–53.
49. See Keith Oatley, *A Brief History of Emotions* (Oxford: Basil Blackwell, 2004), 70–1.
50. Sidney Ochs, *A History of Nerve Functions: From Animal Spirits to Molecular Mechanisms* (Cambridge: Cambridge University Press, 2004), 112.
51. William B. Carpenter, *Principles of Mental Physiology* (New York: Appleton, 1874), 14.
52. Ochs, 108; I am summarizing here from Ochs's more detailed account of this history, 108–21. See also Edwin Clarke and L.S. Jacyna, *Nineteenth-Century Origins of Neuroscientific Concepts* (Berkeley: University of California Press, 1987) who suggest that the "history of concepts of nerve function is one of the longest in the evolution of neurosciences" (157). They classify the first period as preceding Galvani (1791); the second dates from 1791–1840, and the third from Emil du Bois-Reymond in the 1840s to the present day.
53. *Researches in Animal Electricity* was published over many years, the first part appearing in 1848, the last in 1884. Clarke and Jacyna, *Nineteenth-Century Origins of Neuroscientific Concepts*, credit Du Bois-Reymond with placing the discipline of electrophysiology of nerve and muscle on a firm scientific basis (157). See also Kurt Danziger, *Naming the Mind: How Psychology Found its Language* (London: Sage, 1997), 62.
54. Alexander Bain, *Mind and Body*, 36–7.
55. Edwin Morris, *A Practical Treatise on Shock* (1867; Philadelphia: J.B. Lippincott, 1868), 18.
56. Arrhenius's theory was that even in the absence of an electric current flowing through a solution, electrolytes (substances able to dissolve in water to yield

a solution that conducts electricity) can dissociate into ions, which are electrically charged particles.

57. See Danziger, *Naming the Mind*, 39–41.
58. Thomas Dixon, *From Passions to Emotions: The Creation of a Secular Psychological Category* (Cambridge: Cambridge University Press, 2003), 3.
59. Dixon, *From Passions to Emotions*, 4.
60. See Dixon, *From Passions to Emotions*, 143–4.
61. Bain, *Mind and Body*, 132. See Dixon's discussion of this passage, *From Passions to Emotions*, 144.
62. Dixon and others argue that this tendency reaches its apotheosis in William James, well known for inverting prevailing assumptions and suggesting that we are sad because we cry rather than vice versa.
63. Dixon, *From Passions to Emotions*, 24. Whereas he sees the dual-aspect monism of Bain and others as a tacit epiphenomenalism, I think its distinction from Huxley's position is worth preserving. Significantly, Dixon does not discuss the influential work of William Carpenter in this regard.
64. Bain, *Mind and Body*, 41.
65. Menke, "Fiction as Vivisection," 623.
66. George Eliot, *Middlemarch*, 39: 382–3. See Michael Davis, *George Eliot and Nineteenth-Century Psychology: Exploring the Unmapped Country* (Aldershot: Ashgate, 2006), which came out just as this chapter was going to press, precluding a fuller engagement with Davis's work; he discusses this and other passages in which George Eliot uses electricity to analogize internal processes (151–2).
67. Martha N. Nussbaum, *Upheavals of Thought: The Intelligence of Emotions* (Cambridge: Cambridge University Press, 2001), 25.
68. See for example Joseph LeDoux, *The Emotional Brain, the Mysterious Underpinnings of Emotional Life* (New York: Simon and Schuster, 1996); and Jaak Panksepp, *Affective Neuroscience : The Foundations of Human and Animal Emotions* (Oxford: Oxford University Press, 1998); and Paul Thagard, *Hot Thought: Mechanisms and Applications of Emotional Cognition* (Cambridge, Massachusetts: MIT Press, 2006).
69. See Roz Carroll, "'Thinking through the Body': An Interview with Allan Schore," http://www.thinkbody.co.uk/papers/interview-with-allan-s.htm (accessed January 21, 2007); and Allan N. Schore, *Affect Regulation and the Repair of the Self* (New York: Norton, 2003).
70. See Nicholas Dames, *Amnesiac Selves: Nostalgia, Forgetting and British Fiction 1810–1870* (Oxford: Oxford University Press, 2001); on Victorian theories of memory in relation to psychic shock see Jill L. Matus, "Trauma, Memory and Railway Disaster: The Dickensian Connection," *Victorian Studies* 43.3 (2001), 413–36.
71. See for example José Brunner, "Identifications, Suspicions, and the History of Traumatic Disorders," *Harvard Review of Psychiatry* 10.3 (2002),179–84, who notes that "the symptoms of late nineteenth-century disorders encompassed a wide range of phenomena from emotional instability, hallucinations, hyperarousal, physical and mental numbness, auditory disturbances, 'tunnel vision,' and partial paralyses to trembling, nightmares, loss of concentration, persistent headaches, and compulsive vomiting" (179).

8

Medical and Literary Discourses of Trauma in the Age of the American Civil War

Mark S. Micale

More and more, trauma and modernity emerge as mutually constitutive categories. A whole range of Modernist cultural practices and achievements, no less than modern warfare, technology, and industry, as well as modern social and political developments, have entailed "the shock of the new."[1] As humanities scholars take stock of this fact, they converge on the insight that trauma is not timeless and transcendental but culturally and historically produced. The new histories of traumatic modernity are typically set in Europe: they begin (as Jill Matus explains in this volume) with railway disasters in Victorian Britain, move on to the clinics and consulting rooms of Paris and Vienna, pause in the early welfare states of central Europe, flourish in the mass mortality of the First World War, and then culminate in the horrors of the Holocaust. But this genealogy, which elsewhere I have helped to construct, is incomplete, in part because it leaves out America, which is obviously unacceptable.[2] This chapter examines the medical discussion of trauma that was generated by physicians during and after the American Civil War of 1861–65, a militarized setting preceding the 1914–18 war by nearly half a century. Culturally and conceptually, two features mark this early episode: first, while fully formed empirically, American Civil War era writing about traumatic nerve injuries, in comparison with what follows, lacks key components, which makes it at most a kind of pre-discourse of trauma. Second, the interplay of medicine and creative literature, which lies at the center of the present volume, is in full play with this historical story, which has as its most important character a neurologist-novelist whose dual identities informed and fertilized one another.

The classical Greek word "trauma" means "wound," and the adjective "traumatikos" means "pertaining to wounds." The earliest entry for "trauma" in the *Oxford English Dictionary* (*OED*) is from a medical dictionary published in 1693, where it is defined as "a wound from an external cause." According to the *OED*, a 1656 publication cites the adjectival "traumatick," defined as "belonging to wounds or to the cure of wounds." As these definitions indicate, the word's original connotations were strictly physical: in popular parlance during the early modern period, a trauma referred to an actual blow or injury to the body. In eighteenth-century European medicine, the term by extension came to denote the material-pathological effects from such an external assault, that is to say, a pain of the body and suffering following a bodily assault. The hurtful effects of sudden or grave violence on an individual's state of *mind* were occasionally noted, too, but only secondarily. Consequently, medical commentary about trauma during this period was mostly produced by surgeons and military doctors.

Beginning in the last third of the nineteenth century, however, a second line of linguistic evolution within medicine appeared. The medical-military concept of trauma began to be used by clinical observers to describe not only physical but also *mental, cognitive,* and *emotional disturbances* that could result from *either* a bodily *or* an emotional shock. The idea of mental trauma, as *both cause and effect,* began to emerge in the work of Herbert Page in Britain, Jean-Martin Charcot in France, and Sigmund Freud in Austria. By the 1890s, a specifically psychiatric definition was in place, and turn-of-the-century Euro-American psychology is replete with references to "psychical traumata." This range of meanings—trauma as blow or strike, as general pathological condition of the body, and as subsequent mental state—remains available today. It is the key transition over a century ago to psychological meanings of the term that interests me here. This extraordinary linguistic and conceptual mutation, and the wider psychologization of medical understanding that it implies, was a precondition for the formulation of twentieth-century psychogenic theories of traumatic experience.

Military contexts have always been critically important in shaping popular and medical views on trauma. The American Civil War of 1861–65 in particular provides a heretofore unappreciated background to the emergence of the first modern notions of traumatic neurosis. For Union and Confederate soldiers, the daily conditions of life during the Civil War were, in a word, extremely unnerving. Some 620,000 men died during the four-year conflict, a higher number of deaths than in any American war before or since. Thirty-five percent of all fighting soldiers in the Civil War were killed or wounded. The majority of the soldiers were young. They came from civilian backgrounds, and before the war they had rarely been separated from their homes and families.

Soldiers on both sides expected that the outcome of the rebellion would be quickly decided in a few local skirmishes and were wholly unprepared for the four-year bloodbath. In the huge corpus of Civil War letters, memoirs, and autobiographies, two complaints recur during the first two years of the conflict: the grinding boredom of camp life and the physical exhaustion from long marches, often 10 to 20 miles a day, at times in inclement weather. As the war dragged on, shortages of food and clothing became an increasing problem as well.

To these conditions were added the dire threats of disease. Sanitary illnesses spread quickly in camp life. Infectious diseases, foremostly cholera, typhoid, malaria, smallpox, measles, mumps, tuberculosis, and dysentery, became chronic scourges. For every battle death, two men died of disease—an estimated total of about 164,000 Confederates and 250,000 Federals. What is more, this was a period when medicine practiced neither antisepsis nor anesthesia. In many instances, military surgeons contributed to the fatality of wounds by amputating wounded limbs, probing uncleaned wounds with dirty instruments, and binding wounds tight with unsterilized bandages. The medical historian Paul Steiner estimates that under these conditions over 30,000 shock-inducing amputations were performed on Union soldiers, a fearsome 28% of which resulted in the soldier's death. The prevalence of disease among fellow soldiers; the sights, sounds, and smells of army "hospitals" near the fighting front; the knowledge of exorbitant casualty rates; the presence of armless and legless soldiers; and the burial of large numbers of war dead could only have been deeply disturbing to the troops.

Worst of all of course was the battle experience itself. From the Firing on Fort Sumter on April 12, 1861, signaling the start of the war, to General Lee's surrender at the Appomattox Courthouse on April 9, 1865, the loss of life in the Civil War was grotesquely greater than anyone had anticipated. It was not just that war is always more horrific in reality than in rhetoric. It was, rather, that advances in the technology of war were capable of inflicting new types and intensities of destruction on soldiers' bodies and minds. The military historian Richard Gabriel has observed that

> it was on the battlefields of Antietam, Gettysburg, and Chancellorsville that troops made their first headlong frontal assaults against repeating rifles and pistols whose rates of fire caused thousands of casualties. . . . The use of the delayed timed artillery fuse which allowed artillery rounds to burst above the heads of advancing soldiers increased the number of head wounds dramatically. The lack of any protective

> headgear also contributed to the number of head and neurological injuries. The Gatling gun, a somewhat basic but nonetheless effective machine gun, [also] made its appearance with devastating impact.[3]

Modern industrialized weaponry, in short, made the Civil War battlefield far more lethal than it had ever been before. Especially from the Battle of Chancellorsville in May 1863 onward, the military nature of the war changed fundamentally. The historian John Talbott has noted that the material fighting conditions of the last year of the Civil War—exposure to concentrated doses of artillery fire; entrenched defenses and siege warfare; mechanized violence administered on a massive scale; and an impersonal and technologized relation between soldiers—strikingly foreshadow the circumstances of twentieth-century wars.[4]

In light of these new fighting conditions and destructive realities, it is not surprising that cultural representations of the war diverged sharply from those of earlier wars. Literary historians have long remarked that the U.S. Civil War gave rise to no heroic epics along the lines of the novels of Tolstoy and Stendhal about the Napoleonic wars. In place of grand literary narratives, the war generated a remarkable volume of diaries, private letters, and personal memoirs chock full of descriptions of death and suffering. Artistically, Stephen Crane's *The Red Badge of Courage* (1895) is widely judged the finest novel about the war. Yet, indicatively, Crane de-glorifies his subject, describing the fear, chaos, and confusion of the battlefield, and he offers a rare account of a soldier's masculine identity in crisis. Along the same lines, the most famous pictorial images of the war are Mathew Brady's powerful and appalling photographs of corpse-strewn battlefields. In its questioning of traditional war romanticism, the Civil War again casts forward to our more present-day attitudes about war in a nuclear age. It is no coincidence that the most enduring military *bon mot* to emerge from the conflict was General Sherman's "War is hell."[5]

Strikingly, the advent of modern American warfare in the 1860s brought with it the first documentary evidence of what today would be regarded as traumatic neurosis on a large scale. Recent Civil War scholars have discovered that the medical, autobiographical, and epistolary writings of the Civil War are in fact rich in powerful and poignant accounts of acute, war-related mental and emotional suffering.

In these accounts, fear, terror, anxiety, and anger mix with the exhilaration and exhaustion of fighting. At Gettysburg, one Confederate soldier recalled the scene of thousands of his comrades "about to face death and the awful shock of battle."[6] Another Union soldier, Private Wilbur

Fisk, registered in a letter dated May 9, 1864 "the terrible nervous exhaustion of fighting."[7] Officers were not spared from these painful experiences. The future Supreme Court Justice Oliver Wendell Holmes Jr. was a member of the 20th Massachusetts Regiment, which had engaged in much hand-to-hand fighting since the autumn of 1861, and Holmes himself was wounded three times. Near his breaking point, young Holmes on June 24, 1864 wrote from the killing fields of northern Virginia to his parents back home:

> These last few days have been very bad. Many a man has gone crazy since this campaign begun [*sic*] from the terrible pressures on mind and body. . . . I hope to pull through but I don't know. Doubt demoralizes me as it does any nervous man. I cannot now endure the labors and hardships of the line.[8]

From the same regiment, Captain H.L. Patten reckoned that his men "have been so horribly worked and badgered that they are utterly unnerved and demoralized." Capturing the corrosive psychological effect of prolonged fear and anxiety, Patten added of his comrades that "[T]hey are easily scared as a timid child at night."[9]

Particularly disturbing to soldiers was the spectacle of fellow soldiers dead, dying, or mutilated on the battlefield. A Rhode Island chaplain found these sights "appalling to overtaxed nerves" while a young Virginian said it was a sight "to make the blood run cold in the stoutest heart."[10] Similarly, an Ohio soldier recalled the terrifying atmosphere at the Battle of Franklin:

> The booming of cannon, the bursting of bombs, the rattle of musketry, the shrieking of shells, the whizzing of bullets . . . the falling of men in their struggle for victory, all made a scene of surpassing terror and awful grandeur.[11]

Most harrowing were the frontal assaults against a wall of deadly musketry and artillery fire, typified famously by Confederate General George Pickett's suicidal charge at Gettysburg. One survivor recalled graphically "[b]rains, fractured skulls, broken arms and legs, and the human form mangled in every conceivable and inconceivable manner. . . . At every step they take they see the piles of wounded and slain, and their feet are slipping in the blood and brains of their comrade."[12] "The half can never be told," another Northerner bemoaned. "Language is all too tame to convey the horror and the meaning of it all."[13]

In twentieth-century warfare, psychiatric breakdown is especially associated with exposure to concentrated artillery fire on stationary positions, of the sort experienced by British, French, and German infantry soldiers along the Western front during World War One or by American and Japanese soldiers in the South Pacific islands during World War Two. On this point, the last year of the Civil War yet again anticipates the twentieth century. The new trench system's effect on the mind was vividly expressed by the essayist and novelist John W. DeForest:

> The nuisance of trench duty does not consist in the overwhelming amount of danger at any particular moment, but in the fact that danger is always present. The spring is always bent; the nerves never have a chance to recuperate; the elasticity of courage is slowly worn out . . . the health is enfeebled by confinement, and the nervous system shaken by incessant peril.[14]

This combination of circumstances—the destructive power of artillery shells, the passive vulnerability of stationary soldiers, and the excruciating wait for launched shell to explode—seems to have been uniquely destructive of soldiers' psychic defenses.

Neither was fear of imminent death in battle the only source of war-induced stress and strain. What comes through clearly in countless soldiers' letters, officers' diaries, physicians' recollections, and medical case records is the number and diversity of experiences that could undermine the mental and emotional hygiene of the troops. The rigors of training, the fatigue of marching, the anxiety of immobility, the anticipation of combat, the burdens of command, the fear of enemy imprisonment, the guilt of surviving, and the moral trauma of killing all contributed to stress and hardships of soldiering.

What sort of language did contemporaries—both medical and lay persons—use to describe and diagnose these experiences? No psychiatrists—in any past or present definition of the term—played a role in the Civil War. We look in vain for full descriptions or formal diagnoses of post-traumatic psychopathology. Recently, however, a number of historical scholars have established convincingly that, camouflaged under a variety of other labels are unmistakable signs of the devastating and often enduring psychological consequences of the war.

For centuries, "nostalgia" was a diagnostic category in European medicine that designated a morbid and excessive dwelling on thoughts of hearth and home. First formulated by a German medical student in the late-seventeenth century, and then widely applied by French doctors during the Napoleonic campaign when mass conscription to wars in foreign lands occurred, the medical concept of nostalgia was revived in American military medicine during the 1860s and 1870s.[15] According to *The Medical and Surgical History of the War of the Rebellion*, the massive, government-sponsored account of the war, 5,213 Federal troops were treated or discharged for "nostalgia" during the first year of the war alone. Rates rose somewhat in the later years. Nostalgia encompassed a wide range of behaviors from common homesickness to severe pathological longing that could end in death.[16] A form of intense separation anxiety, its most common symptoms were fatigue, sleeplessness, loss of appetite, heart palpitations, fevers, anxiety, and depression. New, young recruits from rural communities serving far from home were particularly prone to the affliction. Its potential mortality, which allegedly could be allayed only by returning a soldier home, was widely acknowledged.

Two other relevant categories are desertion and malingering. Over 300,000 Union and Confederate soldiers deserted from their ranks during the Civil War, many of whom returned to service, and many of whom did not. The images of terrified infantrymen flocking to the rear and glassy-eyed soldiers wandering in a deranged state after a battle are memorably depicted in Crane's *Red Badge of Courage*. Along similar lines, so-called malingering became an important category in Civil War medicine. Among officers and physicians there was a strong suspicion that soldiers presenting mental or physical symptoms without evident bodily injury were simulators trying to escape service—"shirkers," they were called. Military-medical manuals often included chapters on "feigned diseases," listing among the symptoms lameness, loss of voice or sight, paralysis of the limbs, contracture of the joints, and convulsive fits. *The Medical and Surgical History of the War of the Rebellion* records exceptionally high rates of these symptoms. Army doctors practiced a series of tests, including etherization, to detect cases they believed were frauds. There seems to have been little sense at the time that many cases so classified may have involved physical manifestations of acute psychological distress.

Likewise, Eric Dean has noted that many of the most common and debilitating ailments experienced by Civil War infantrymen, such as headache, exhaustion, vomiting, gastric upset, and chronic diarrhea, were nonspecific symptoms that could well be stress related. "[C]onditions during the Civil War were so wretched," Dean explains, " . . . that it is often impossible to tell if the men were suffering from a specific somatic illness, overall exhaustion, express psychological breakdown, or a mixture of all these conditions."[17] The fact that soldiers were granted relief from duty only if they revealed a clearly physical symptomatology may well have been an added incentive to somaticize psychological sickness.

"Insanity" is a final diagnosis deployed in the historical sources. The only category of psychological illness recognized by medics, officers, and the government during the Civil War was out-and-out madness. According to the medical historian Albert Deutsch, 6% of the medical discharges from the Union Army were granted to soldiers deemed insane. Eric Dean states that approximately 1,230 Union soldiers were committed to the recently created Government Hospital for the Insane, as St. Elizabeth's Hospital in Washington, D.C. was then known. The hospital Superintendent at that time, Dr. Charles Nichols, commented in an annual hospital report that by the war's end, mentally deranged soldiers accounted for 83% of his total admissions. Many other soldiers were sent to local asylums, or were released to relatives, or were just discharged from the army without supervision.[18] In mid-nineteenth-century American medicine, the omnibus category of insanity divided into the large subcategories of mania, melancholia, and dementia. Dean finds, however, that records at state asylums often include more specific descriptions of these soldier-patients' condition, such as "War Excitement," "Exposure in the Army," "Shock of Battle," or simply "The War."

In addition, Dean has gathered highly suggestive evidence about the morbid mental condition of many Civil War Union veterans. He presents dozens of instances of spouses, children, and neighbors who observed agitated, eccentric, or morose behavior in soldiers returning from the war. He documents cases of insomnia, nightmares, trembling, and crying spells among veterans many years after the conflict had ended. Examining the medical files of state asylums as well as federal disability pension records, he has discovered hundreds of soldiers who, long after the silencing of the guns at Appomattox, collapsed mentally and required institutionalization. Among inmates at the Indiana State Asylum during the period 1861 to 1920, for example, Dean locates 134 cases involving Union veterans who suffered from "mental derangement," "shell concussion," "unsound mind," "affection of the head," "brain trouble," "nervousness," and "nervous prostration." He also notes an exceptionally high degree of alcoholism as well as addiction to opium, cocaine, and morphine among both Confederate and Union veterans. "Many of these men continued to suffer from the after-effects of the war and, along with their families, often lived in a kind of private hell involving physical pain, the torment of fear, and memories of killing and death."[19]

The most detailed case that historians have unearthed is that of Newell Gleason. A successful engineer before the war, Gleason was commissioned lieutenant colonel with the 87th Indiana Regiment, serving from the summer of 1862 until the end of the war. His unit fought in battles at Chattanooga and Atlanta. Gleason distinguished himself with bravery during these fierce clashes; but soon afterward his men began to notice in him strange behaviors, including wild swings in mood from despondency to jubilation. During a trip home to Indiana with a physician-friend, Gleason became frantic and had to be sedated chemically. After the war, he tried to resume his engineering career but found that he was unable to concentrate. His condition deteriorated, and in 1874, during a three-day episode of "violent mania," he had to be temporarily committed to the Indiana State Hospital for the Insane. Periodically his symptoms remitted, but his mental anguish continued over the years. In 1886, he threw himself headlong down the cellar stairs of his home and died of a fractured skull. The coroner judged Gleason's death a suicide; but another physician concluded that Gleason's condition had been "induced by disorders and hardships incurred in the U.S. Military Service"—some 22 years earlier![20]

Of the combat soldiers' inner experience of these events, we know almost nothing today. The passages cited above rarely consist of more than a sentence or two of description. Conspicuously absent in the stories told by Civil War soldiers about themselves is the impulse to confession and self-revelation that has become so pronounced in our own day. Rather, their accounts draw on a language of suffering that is plain, nontechnical, and notably nonpsychiatric. Soldiers and officers alike speak of feeling sad, melancholy, overwhelmed, demoralized, dispirited, or disheartened, of having the blues, of being nervous, of being worn out or played out. They briefly describe their feelings and experiences, but they do not conceive of themselves as victims of a specific disease or syndrome. During the American Civil War there was no official category comparable to such later diagnoses as "shell shock" (World War One), "combat neurosis" (World War Two), or Post-Traumatic Stress Disorder (Vietnam and after).

Attitudes among military authorities were at best unsympathetic and uncomprehending; at times, they were brutal and sadistic. As Talbott and Dean have noted, symptoms of stress in the absence of some physical wound or organic illness were often viewed as cowardice or as an attempt to shirk duty. Officers typically interpreted these cases in the explicitly moral language of weakness and lack of bravery, and their official responses were often judicial and disciplinary. It is impossible to know how many of the men who were discharged dishonorably, sent to military prisons, court-martialed, or shot as deserters today would be viewed as psychiatric victims of war.

And how were these symptoms and behaviors classified, conceptualized, and treated by physicians, which is to say by the community that was professionally trained and culturally sanctioned to heal the diseased and console the suffering? Civil War doctors tended to employ a number of informal labels to account for these cases. Combining popular and medical terminology, they diagnosed "sunstroke," "shock of battle," "nervous debility," "irritable heart," and "war excitement." Interestingly, in the closing year of the war the new term "shell explosion" began to appear in medical records. Physicians used these terms, however, to describe symptoms, not to designate a discrete disease entity. Furthermore, for virtually all these cases, including those thought to be insanity, Civil War doctors held a strictly physicalist explanation. They spoke of strained nerves, sick brains, and damaged spinal columns, but not of suffering psyches. They routinely speculated about a hereditary inclination to mental breakdown and pondered the role of a nervous or melancholic temperament. Since there was no theory or nomenclature of traumatic neurosis in 1860s American medicine, it is not surprising that therapeutics were nearly nonexistent. Many military physicians believed that a return to battle served most effectively to get a soldier's mind off his problems. Confronted with more severe cases, army doctors were by and large helpless before a mysterious phenomenon. The only treatments that seemed to work were rest, relief from the fighting line, and, at times, a furlough. Their only prescribed medicines were tonics and sleep-inducing agents.

One group of American military physicians in the Civil War era was especially well-qualified to appreciate the complex significance of these cases. In August 1862, the brilliant and controversial Surgeon General William Hammond created a new hospital on Turner Lane in Philadelphia, which was then arguably the leading medical city in the United States. Hammond established the new facility in order to treat the flood of cases of nerve wounds, caused by gunshot and shrapnel fire, occurring on northern battlefields. The Turner Lane clinic housed 400 beds, and hospitals across eastern Pennsylvania transferred their nervous and neurological cases to its wards. Hammond appointed a remarkable quartet of young medical talent to staff the clinic: the physiologist and neurologist Silas Weir Mitchell, the surgeon William Williams Keen, the general practitioner George Read Morehouse, and the cardiologist Jacob Mendez Da Costa.

As Mitchell later wrote, the war poured into the Turner Lane clinic "a continued flow of damaged humanity;" and by 1865, he and his team were "well lessened [*sic*] in the terrible school of war."[21] Their most bracing experience came in early July, 1863 with the three-day Battle of Gettysburg, 80 miles west of Philadelphia, where 23,000 Union soldiers and 28,000 Confederate soldiers lost their lives. Fifty years after his visit to Gettysburg, Mitchell was still haunted by "the dead Confederates lying below Round Top the day after the fight, with arms and legs in rigid extension" and "that strange complex odour which rises from a battlefield." "A most horrible memory," he added.[22]

"Perhaps the outstanding medical event of the Civil War," the medical historian Albert Deutsch once remarked, "was the rise of the neurological profession in America."[23] In no small part, this achievement was due to the combined work of Mitchell, Keen, Morehouse, and Da Costa. The Turner Lane facility was in effect the first clinic for the study and treatment of nervous disorders and neurological injuries in America, and it was operating under wartime circumstances. "The cases were of amazing interest. Here at one time were eighty epileptics, every kind of nerve-wound, palsies, singular choreas, and stump disorders. Never before in medical history," Mitchell recalled somewhat exaggeratedly, "has there been collected for study and treatment so remarkable a series of nerve injuries."[24]

The first fruit of this extraordinary clinical opportunity was *Gunshot Wounds and Other Injuries to the Nerves*, published by Mitchell, Morehouse, and Keen. A classic of early American neurology, *Gunshot Wounds and Other Injuries to the Nerves* consists of 43 vivid case reports of large nerve wounds in soldiers.[25] The book advanced medical knowledge of traumatic neuralgias; primary and secondary neurological shock; the neurological aspects of heatstroke, chorea, and meningitis; and the symptomatology and treatment of peripheral nerve injuries. Throughout the work, the authors seem aware of the possible *psychological* dimensions to many of their cases. They describe baffling cases in which headaches, tremors, hearing loss, neuralgias, and gastrointestinal upset develop independently of a bodily injury, and they record the persistence of quasi-neurological symptoms long after the healing of a tissue wound.

Along much the same lines, Mitchell in 1871 independently published a remarkable study of "phantom limbs."[26] This was a curious type of case in which soldiers with amputated legs or arms experienced sensations of pain or numbness in the lost extremity. In another work, Mitchell, Keen, and Morehouse together explored "reflex paralyses," or cases involving the localized loss of sensation and movement in body parts far removed from a physical wound. The three men also gathered thousands of pages of clinical observations about convulsive cases, although fire destroyed their notes and prevented publication; but Mitchell later recalled some of their intriguing clinical observations, including cases in which they were able, by means of localized stimulation, to elicit an epileptoid seizure in soldiers who had previously undergone a physical or mental shock. These cases, Mitchell added in retrospect, "would have entirely anticipated the delineation of those disorders which we now accept under the name neurasthenia."[27]

Reading *Gunshot Wounds and Other Injuries to the Nerves*, a New York army surgeon sought to highlight "an additional variety in the list of injuries to the nervous system."[28] In the article "Cases of Injuries to the Nervous Centres, from Explosion of Shells, without Wound or Contusion," also published in 1864, Dr. George Burr brought to light three cases of soldiers who were stunned but physically unharmed by the explosion of artillery shells close to them on the battlefield. In one of Burr's cases, Captain R. P. W. experienced paralysis of an entire side of his body as well as his tongue and neck muscles following an explosion during an attack at Charles City Cross Roads in Virginia. In an autobiographical excerpt, another of Burr's shell-shocked patients describes his loss of energy, creeping paralysis, inability to speak, and sensations of pain and tenderness across the skin following an explosion. "In neither of the preceding cases," Burr insists, "was there wound or contusion."[29] Burr's article might well have been drawn from the medical literature of World War One on shell shock.

Perhaps the most clinically significant piece of work to emerge from the Turner Lane clinic was conducted by Jacob Mendez Da Costa. Whereas Mitchell, Keen, and Morehouse ranged widely in their publications, Da Costa concentrated on a single symptom category. Among his soldier-patients, Da Costa repeatedly observed a strange phenomenon: cardiac complaints in the absence of any detectable physical pathology. In 1871, Da Costa published "On Irritable Heart: A Clinical Study of a Form of Functional Cardiac Disorder and Its Consequences" in the *American Journal of the Medical Sciences*. The article has since become a minor medical classic. Da Costa presented over 300 cases of infantry soldiers who during and after the war suffered a range of cardiovascular symptoms, including burning or stabbing chest pains, altered pulse, nervous palpitations, and rapid or panicked breathing. Exhaustive examination of these patients by Da Costa revealed no structural abnormalities of the heart or signs of physical disease. In many instances,

Da Costa observed, the complaints were sufficiently severe to incapacitate the soldier. Moreover, the ailments often occurred in soldiers who seemed nervous or distraught, and they were sometimes accompanied by headaches, insomnia, or nightmares. In nearly all cases, the symptoms vanished mysteriously after the war. Da Costa labeled the syndrome "irritable heart."[30]

Da Costa's concept quickly entered the medical mainstream. In the 1880s, the American neurologist George Beard included irritable heart among the symptoms of neurasthenia, as did Sigmund Freud in his description of "anxiety neurosis" in the 1890s. During the Boer War and World War One, British physicians investigated cases of "soldier's heart" and "cardiac neurosis." In the early 1940s, confronted with yet another wartime outbreak of such symptoms, physicians re-dubbed these cases "Da Costa's Syndrome," as indeed they are known today.

The articles and monographs of Mitchell, Keen, Morehouse, Burr, and Da Costa written during the 1860s and 1870s are brilliant, pioneering documents in the history of American neurology and cardiology. Conspicuously, however, these writings did not contribute to the *psychiatric knowledge* of their (or our) day. In light of the great volume and variety of stricken soldiers that Mitchell's team encountered, we might well expect the first American theory of traumatic neurosis to have emerged from their wartime work in Philadelphia (as it would shortly begin to emerge in London, Paris, and Vienna). To the early twentieth-first-century reader, their cases cry out for psychological understanding.

Yet Mitchell and his colleagues never probed, either clinically or theoretically, the psychology of these cases. To the contrary, their clinical analyses remained doggedly somaticist: in *Gunshot Wounds and Other Injuries to the Nerves*, the authors' attention is fixed on the original physical injury; they did not explore the possible role of mind and emotion in shaping, prolonging, or embellishing these illnesses. In their discussions of convulsions in soldiers, we search in vain for any notion of the possible intermixture of epileptic and hysterical elements. And, for the most inexplicable symptom formations, Mitchell et al. lapsed into the value-laden vocabulary of malingering and feigned diseases.[31] Correspondingly, in his presentation of nerve injuries caused by shell explosions at a distance, Burr speculates that an unknown medium in the atmosphere injured the soldiers' nervous systems. Da Costa likewise proposed that his "exhausted hearts" traced to hard marching, physical overexertion, and poor nutrition. Mitchell, Kean, Morehouse, Burr, and Da Costa often mention in passing a patient's combat experience or record his state of mind. Emphatically, however, they give no sign of understanding that mental and emotional shock could act as *a primary causative force* in these cases; nor did they seem to perceive that fear, anxiety, and depression could manifest themselves in direct and complex ways through the language of the body.

Just how remarkable a lost scientific opportunity this was comes more sharply into focus if we consider Silas Weir Mitchell's famous later career. After completing his work on the neurology of war, Mitchell turned exclusively to private medical practice. Specifically, he concentrated on treating well-to-do adult women and adolescent girls with all manner of nervous complaints, a social world he was introduced to through his second marriage. Based on his work with this new and lucrative clientele, he published *Wear and Tear* (1871) and *Fat and Blood* (1877). It is in these highly popular books that Mitchell propounds his famous "rest cure" for nervous diseases, a regimen of rest, massage, overfeeding, and mild electrotherapy. Both of these works, as well as his *Lectures on Diseases of the Nervous System, Especially in Women* (1881), present dozens of cases, nearly all involving women and children, despite the fact that Mitchell treated adult men in his private medical praxis.[32] In these later works, Mitchell is less concerned with physical wounds of the nerves than with "the functional nervous disorders," including hysteria, neurasthenia, and nerve prostration, three terms he then began using regularly in his civilian practice. Furthermore, in his published case histories of the 1870s and 1880s, Mitchell routinely discusses purely emotional events—family deaths, romantic frustration, and financial misfortune—as the causes of nervous breakdown.

Today, this is the stage of his career that is best known: Mitchell as manipulative and paternalistic caretaker of such gifted but vulnerable women as Charlotte Perkins Gilman and Edith Wharton. From our perspective, what is surely noteworthy is that the figure who became the most prominent "nerve doctor" of Gilded Age American society began his career as an army doctor treating soldiers fresh from the battlefield. Mitchell later acknowledged that he first formulated the idea of rest as a curative agent in his earlier treatment of nerve-injured Civil War soldiers.[33]

On one occasion, Mitchell brought together very briefly his work on female hysteria in civilian society and nervous disorders in Civil War soldiers. The 10-page essay "Clinical Lecture on Nervousness in the Male," published in 1877, is an intriguing transitional text between his early and later medical careers. Nervousness, Mitchell claims at the essay's outset, is usually thought "entirely the sad prerogative of woman."[34] He intends, however, to report on a category of patient that he has encountered in his work at a local public clinic but that had rarely been remarked upon by doctors: the *working-class male neurotic*.

Mitchell discusses seven such cases, one of which deals with a war veteran:

> C.J., age 30, a large vigorous man, was one of the leaders of the gallant storming party which took Fort Fisher. He had before this led a life of adventure and danger. Probably it would have been hard to find a more cool, quiet, resolute soldier. He was shot through the right lung, and the ball emerging entered the arm and injured the median nerve.[35]

C.J. recovered from the tissue injury. However, the experience, Mitchell continues,

> transformed him [the patient] morally into a peevish, timid child. In this condition he was a type of a nervous man. A sudden noise, an unlooked for letter, the morning visit of the doctor, set his heart beating and made his limbs cold. At the prospect of a hypodermic injection he would shed tears, and he became annoyingly susceptible to the irritative influence of sounds and light.[36]

Mitchell goes on to reflect on the possible pathological effects of fear and anxiety. He classifies C.J. among the "cases of sudden nervousness . . . which arise from mental shock," adding that "(h)ow the extreme of nervousness acts on the battle field is well known."[37] In all but name, he diagnoses this ex-soldier with "traumatic hysteria," as Charcot in Paris will label it a decade later. Mitchell does not pursue his line of observation further, however, and to the best of my knowledge this is the only case history of its kind that he ever published.[38] Conspicuously absent from his postbellum medical output are the nerve-damaged veterans, largely from rural and working-class backgrounds, whom Mitchell treated in such great numbers during the war.

Finally, Mitchell's nonmedical writings are also germane. During the second half of his career, Mitchell's fame as a man of letters equaled his reputation as a physician. Between 1866 and 1913, working in the mold of the nineteenth-century physician-litterateur, Mitchell wrote some 15 novels, several of which reached national reading audiences in multiple editions. His collected fiction and poetry appeared in a 12-volume edition in 1909.[39] At least five of Mitchells novels—*In War Time* (1884), *Roland Blake* (1886), *Circumstance* (1901), *Constance Trescot* (1905), and *Westways* (1913)—deal, centrally or peripherally, with the Civil War, including frontline and home front settings. Artistically, Mitchell's creative fiction is not judged highly today. What interests us is that in his day, specifically through his novel writing, Mitchell was regarded as a keen scientific student of the human personality. Readers at the time repeatedly cited his clinical-like powers of observing the human personality and the realistic, "diagnosable" quality of his characters' suffering. During the 1950s, appreciative psychoanalytic readers in search of Freudian precursors christened him "a psychiatric novelist."[40]

Most important, as contemporaries also recognized, many of Mitchell's fictional characters bear striking resemblances to patients in his case-historical writings. Mitchell's first published piece of imaginative fiction, which appeared in *Atlantic Monthly* a year after Civil War hostilities had ceased, was "The Case of George Dedlow" (1866).[41] In this intriguing short story, Mitchell depicts an ex-soldier (and army surgeon!) whose four extremities had to be amputated because of horrible injuries yet who continues to experience extravagant sensory hallucinations in his arms and legs—one of the first descriptions of the "phantom limb" concept.[42] Beginning in the mid-1880s, when at the height of his private practice with neurasthenic women, Mitchell regularly began to write full-length novels, several of which featured adult male characters that were nervously debilitated. Lisa A. Long has highlighted the fact that the experience and memories of the Civil War haunt the male protagonist of Mitchell's first such work of fiction, the semi-autobiographical *In War Time* (1884).[43] Likewise, in *John Sherwood, Ironmaster* (1911), Mitchell introduces the character Bob Cairns who is always sick before a battle and is branded a coward by fellow soldiers but who later saves a man's life. Several of Mitchell's novels also include overtly insane characters who were war veterans. In *Constance Trescot*, one John Greyhurst, a southern officer formerly in the Civil War, commits a murder, goes mad, and eventually takes his own life. Similarly, Mitchell's popular second novel, *Roland Blake* (1885), which deals with Grant's Wilderness Campaign, features the Confederate Richard Darnell who after the war becomes a drunkard, loses his fortune, and then dies. At the opening of one of the novel's last chapters, Mitchell suggests an apt analogy, drawn from clinical medicine, to account for Darnell's downfall:

> The surgeon's idea of a "shock" as a result of sudden physical injury should be imported into the domain of criminal psychology . . . In the sphere of mind and morale the abrupt shock of fear or shame may in like manner affect distant nerve-cells and thus deaden memory, palsy the organs of reason, annihilate for a while the power

> to love or hate, and even reduce a man for a time to the verge of inert idiocy. Under such a blow . . . Darnell found himself all day incapable of thought, stupid, and indifferent.[44]

In light of these observations, we might think that Mitchell, working in the freer, more expressive and subjectivist medium of creative fiction, would have probed in greater depth the private world of the soldier's psyche. With its narrative nature and emphasis on individual states of consciousness, imaginative literature arguably is better suited than science, especially during this high positivistic phase, to explore such mental states. But neither of Mitchell's authorial selves—the neurological or the literary—undertakes this sensitive project. The excerpt above is limited to a passing excursus in a single novel. While many of his fictional characters fought in the war, Mitchell never in an explicit, causal way connects their subsequent deterioration to their war experience. And in depicting his mentally ill characters, and explaining their pathology, he typically turns to the age-old discourse of temperament and to the new hereditarian model of madness that was then gaining medical ground.

There is a postscript to Mitchell's tentative position in the history of trauma. In 1913, on the eve of World War One's outbreak in Europe, Mitchell addressed the Physicians Club of Chicago. He was 83 years old and in the last year of his life. Mitchell took the occasion gently to scold a new medical generation for failing adequately to recognize and appreciate the heroic efforts of government physicians during the Civil War half a century earlier. To this end, Mitchell reviews the conditions under which army doctors labored during those nightmarish years and reviews the types of cases they encountered: "Cases of nostalgia, homesickness," he observes, "were serious additions to the peril of wounds and disease." "I regret," he adds, "that no careful study was made of what was in some instances an interesting psychic malady, making men hysteric and incurable except by discharge."[45]

Much had transpired in Western science and medicine between 1865 and 1914. These changes perhaps granted Mitchell some retrospective insight into what he had seen on the battlefield 50 years ago. We cannot help but wonder if Mitchell, had he lived a few years longer, would have recognized a new outbreak of that "interesting psychic malady," a still larger epidemic of male hysteria among young soldiers who this time were fighting in the trenches of northern France and Belgium.

Mitchell's rueful comments near the end of his life reflect his own scientific regrets; but they also reveal the inevitable limitations of his medical generation. Immediately after the end of the war, the large majority of army doctors—Mitchell, Keen, Morehouse, and Da Costa included—turned to private practices in the civilian world. The pioneering Turner Lane clinic in Philadelphia closed. During the 1870s and 1880s, the three thick volumes of the *Surgical and Medical History of the War of the Rebellion* were published, including masses of information about dozens of diseases. But, for all its comprehensiveness, the work is totally silent about psychiatric illnesses among Civil War soldiers. Similarly, nervous and mental disorders are entirely omitted from both *Military Medical and Surgical Essays* (1864), which was prepared for the U.S. Sanitary Commission, and from Louis C. Duncan's *Medical Department of the U.S. Army in the Civil War* (1911). I know of no physician who during the remaining decades of the nineteenth century systematically studied or treated psychologically disturbed Civil War soldiers and veterans. Needless to say, the situation would be very different regarding American wars in the twentieth century.

More than a century before formulation of the diagnosis Post-Traumatic Stress Disorder (PTSD), the American Civil War generated in all likelihood thousands of psychiatric casualties. Why, then, was there no discourse of psychological trauma during and following the war? How many other cases similar to that of S. Weir Mitchell's "C. J." must there have been at the time? And why were Mitchell, Keen, Morehouse, Da Costa and their medical contemporaries unable or unwilling to explore the psychological aspects of the cases they encountered?

I think several overlapping factors were at work. First, a suitable professional infrastructure of psychiatry had not yet developed in the country. In the Civil War era, the only physicians who specialized in the treatment of mental disorders were asylum doctors—alienists, as they were called—who worked in state mental hospitals that housed overtly psychotic patients. General hospitals did not then include psychiatric wards; there was no private-practice psychiatry, no subfield of military psychiatry, and, needless to say, no PTSD clinics. Interestingly, a year into the war a group of asylum superintendents approached the Department of War and volunteered their services on the battlefield, but their offer was rejected. Except occasionally for receiving soldiers discharged from the army as insane, nineteenth-century American "psychiatrists" were absent from the four-year national conflict.

Instead, the large majority of physicians ministering to Civil War soldiers had previously practiced either surgery or general medicine. The rich wartime experiences with nerve wounds gained by these medical men led to the birth of American clinical neurology. But the exclusively organically grounded training of these practitioners caused them to regard mental disorders solely as the result of damaged nerve tissues or disordered brain physiology. Cases not attributable to some somatic condition they tended to interpret as weakness, cowardice, madness, or malingering.

Second, American Civil War medicine lacked the requisite *conceptual apparatus* to interpret cases of psychological breakdown as a distinct category of pathology with an independent etiology. The medical concept of stress had yet to be formulated. Furthermore, no category existed to account for the range of intermediate psychopathological states, between the polar conditions of complete equilibrium and full-fledged madness that would later be designated by the concept of the neuroses. Even more basically, the theoretical foundation for the idea of traumatic psychopathology had not yet been established. Victorian neurological thought, as Anne Stiles points out in this book's Introduction, remained highly materialist. Latter-day psychologies of trauma, however, rest on the belief that emotional and psychological experiences in an individual's environment can operate as primary causative agents. Expressed otherwise, to be recognized and studied, psychological trauma required a theory of psychogenesis in which the mind is credited with a previously unimagined power over the body and the brain. Such a model of mind offers an alternative to interpreting posttraumatic suffering as defective heredity, organic disease, weakness of will, or lack of courage—the four possibilities for the Civil War medical community.

The prominence of epidemic physical disease during the Civil War may well be a third inhibiting factor. In the face of mass mortality from typhoid, dysentery, cholera, and malaria, soldiers' psychological aches and pains probably seemed secondary or insignificant. Physical survival took precedence over mental health. Eric Dean in fact has suggested a kind of inverse relation between the decline of the great infectious killers in modern warfare and the rise of military psychiatry.

These three medical factors—the lack of a suitable psychiatric profession, the absence of key theoretical concepts, and the prevalence of epidemic disease—were reinforced by powerful social and cultural circumstances, which influenced soldiers, officers, physicians, and governments alike. Most apparently, nervous susceptibility in the age of the Civil War evoked the specter of *unmanliness*. In many cultures past and present, aggression generally and war in particular have played a central role in defining conventional masculine identities. War has been widely seen to instill manliness, and an important part of the process of gender socialization was the overcoming of fear with the martial qualities of courage, hardship, and self-discipline. In light of these age-old beliefs, it is likely that Civil War soldiers were reticent about discussing the horrors of the war. To wallow in their emotions would have been considered, by themselves and their society, to be unsoldierly, unmasculine.

Physicians at the time may well have been reluctant to explore these cases for related reasons. Rampant nervous breakdown among soldiers suggested the possibility of widespread "feminine" vulnerability among the very group of adult men in society that was supposed to be the strongest and to protect the homeland. Probing psychological analysis of these cases would almost certainly have required challenging the model of heroic, independent masculinity that underpinned militarist ideology. Victorian-era male physicians, to a degree themselves the product of the same sex/gender regime, may well have lacked the critical distance to undertake such an innovative and sensitive investigation. What is more, such a research project would in all probability have been regarded by contemporaries not as scientifically significant but as indelicate, subversive, even unpatriotic. Not coincidentally, figures like Mitchell found it easier to discuss the functional nervous disorders in women and children—who at the time were of course presumed to be more nervously vulnerable—than in men.

Last of all there was the absence of certain broad cultural circumstances. Widespread social awareness of psychological trauma requires a certain degree of *psychiatric articulateness*. By psychiatric articulateness I mean an ability and a willingness of people, both individually and collectively, to turn inward in order to analyze their feelings, thoughts, and life, to contemplate their emotional states, and to record their psychological experiences in expressive, culturally available forms. In the early twenty-first century, this type of self-awareness and self-revelation is widely viewed as personally desirable, existentially enhancing, and therapeutically beneficial. (Hence the abundant confessional literature of our day as well as countless television revelations about past personal "traumas.")

As we have seen, however, Civil War soldiers groped for a way to make their mental experience of the war comprehensible to themselves, their doctors and officers, their families, and the wider world. As John Talbott has emphasized, the language they turned to was individual, informal, and unclinical. Nervous and mental breakdown among Civil War soldiers involved an acute but vague and inarticulate suffering. It was a personal pain that lacked a name, a diagnosis, a theory, and a therapeutics. In a sense, then, there was no PTSD in the Civil War. In contrast to the American post-Vietnam generations with their proclivity for psychiatric autobiography, the rural and semi-literate populations that largely constituted the Union and Confederate armies 140 years ago mostly remained silent about their experiences during the war. And, as Edmund Wilson pointed out some time ago, American literary intellectuals after the Civil War, unlike the British poets and German novelists of World War One, did not generate a language of memory and suffering in which to record the great shared national tragedy.[46] In all likelihood, the psychiatric history of the Civil War has remained unwritten because contemporaries lacked the appropriate expressive resources for writing it. Not coincidentally, it has taken the experience of later, twentieth-century wars, occurring under very altered medical and cultural circumstances, to alert us today to this hidden dimension of the American Civil War.

Notes

1. Although used previously by other writers, the phrase is associated with the Australian art critic Robert Hughes in *The Shock of the New* (New York: Knopf, 1981).
2. Mark S. Micale and Paul Lerner, eds., *Traumatic Pasts: History, Psychiatry, and Trauma in the Modern Age, 1870–1930* (New York: Cambridge University Press, 2001), chapter 1 and passim.
3. Richard Gabriel, *No More Heroes: Madness & Psychiatry in War* (New York: Hill and Wang, 1987), 106.
4. John Talbott, "'Many a Man Has Gone Crazy': Combat Trauma in the American Civil War," in Talbott, *Mind Wounds: War and Psychic Injury since 1860* (work in progress), chapter 1.
5. "I am tired and sick of war. Its glory is all moonshine. It is only those who have neither fired a shot nor heard the shrieks and groans of the wounded who cry aloud for blood, more vengeance, more desolation. War is hell." William Tecumseh Sherman, Address before graduating class, Michigan Military Academy, June 19, 1879.
6. Spencer Glasgow Welch, *A Confederate Surgeon's Letters to His Wife* (New York: Neale Publishing, 1911), 64.
7. Qtd. in Talbott, "'Many a Man Has Gone Crazy,'" 27.
8. Mark de Wolfe Howe, ed., *Touched with Fire: Civil War Letters of Oliver Wendell Holmes, Jr., 1861–1864* (Cambridge: Harvard University Press, 1946), 149.
9. Qtd. in John Talbott, "Combat Trauma in the American Civil War," *History Today* 46 (March 1996), 45.
10. Qtd. in Eric T. Dean, Jr., "We Will All Be Lost and Destroyed: Post-Traumatic Stress Disorder and the Civil War," *Civil War History* 37 (June 1991), 144.
11. Qtd. in Eric T. Dean, Jr., *Shook Over Hell: Post-Traumatic Stress, Vietnam, and the Civil War* (Cambridge: Harvard University Press, 1997), 57.
12. Qtd. in Dean, *Shook Over Hell*, 58.
13. Qtd. in Dean, *Shook Over Hell*, 57.
14. Qtd. in Talbott, "'Many a Man Has Gone Crazy,'" 35. See also John E. Talbott, "Soldiers, Psychiatrists, and Combat Trauma," *Journal of Interdisciplinary History* 27 (Winter 1997), 437–54.

15. George Rosen, "Nostalgia: A Forgotten Psychological Disorder," *Clio Medica* 10 (1975), 28–51.
16. George Rosen, "Nostalgia"; Klaus Brunnert, *Nostalgie in der Geschichte der Medizin* (Düsseldorf: Triltsch, 1984), passim.
17. Dean, *Shook Over Hell*, 133.
18. William W. Keen, S. Weir Mitchell, and George R. Morehouse, "On Malingering, especially in Regard to Simulation of Diseases of the Nervous System," *American Journal of Medical Science* 48 (1864), 367–94, 377.
19. Dean, *Shook Over Hell*, 100.
20. Qtd. in Dean, *Shook Over Hell*, 153.
21. S. Weir Mitchell, "The Medical Department in the Civil War," *Journal of the American Medical Association* 62 (May 1914), 1447.
22. Cited in Ernest Earnest, *S. Weir Mitchell: Novelist and Physician* (Philadelphia: University of Pennsylvania Press, 1950), 56. As these reminiscences suggest, the war experience took its toll on physicians, too. Today the best known of the Philadelphia group, Mitchell lost his wife to diphtheria and two soldier-brothers during the war years and late in 1864 was compelled to leave Philadelphia for two months of therapeutic recuperation in Europe.
23. Albert Deutsch, "Military Psychiatry: The Civil War, 1861–1865," in J. Hall, G. Zilboorg, and H.A. Bunder, eds., *One Hundred Years of American Psychiatry* (New York: Columbia University Press, 1944), 367.
24. Mitchell, "Medical Department in the Civil War," 1449; Silas Weir Mitchell, *Injuries of Nerves and Their Consequences* (1872; New York: Dover, 1965), 14.
25. S. Weir Mitchell, George R. Morehouse, and William W. Keen, *Gunshot Wounds and Other Injuries of Nerves* (Philadelphia: J.B. Lippincott & Co., 1864). Long admired by neurologists and historians of medicine, *Gunshot Wounds* has recently been newly appreciated and analyzed from a cultural-historical standpoint in Lisa A. Long, *Rehabilitating Bodies: Health, History, and the American Civil War* (Philadelphia: University of Pennsylvania Press, 2004), chapter 1, passim.
26. Mitchell, *Injuries of Nerves and Their Consequences*, chapter 14. Mitchell later published a popular journal article on the subject entitled "Phantom Limbs," *Lippincott's Magazine of Popular Literature and Science* 8 (1871), 563–9.
27. Mitchell, "Medical Department of the Civil War," 1449.
28. George Burr, "Cases of Injuries of the Nervous Centres, from Explosion of Shells, without Wound or Contusion," *New York Medical Journal* 1 (September 1865), 429.
29. Burr, "Cases of Injuries of the Nervous Centres," 431.
30. Jacob Mendez Da Costa, "On Irritable Heart: A Clinical Study of a Form of Functional Cardiac Disorder and Its Consequences," *American Journal of the Medical Sciences* 121 (January 1871), 17–52. See also Paul W. Skerritt, "Anxiety and the Heart: A Historical Review," *Psychological Medicine* 13 (1983), 17–25.
31. Note that their tri-authored study of medical malingering published in the *Philadelphia Medical Journal* appeared in 1864, the same year as *Gunshot Wounds*.
32. On Mitchell's *male* neurasthenic clients, who issued from well-to-do civilian backgrounds, and his rather Rooseveltian remedies for them, see Barbara Will, "The Nervous Origins of the American Western," *American Literature* 70 (1998), 293–316.

33. A point that Nancy Cervetti emphasizes in her excellent article "S. Weir Mitchell Representing 'a hell of pain:' From Civil War to Rest Cure," *Arizona Quarterly* 59 (2003), 69–96. See also Long, *Rehabilitating Bodies*, 30.
34. Silas Weir Mitchell, "Clinical Lecture on Nervousness in the Male," *Medical News* 35 (December 1877), 177.
35. Mitchell, "Clinical Lecture on Nervouess in the Male," 178.
36. Mitchell, "Clinical Lecture on Nervouess in the Male," 178.
37. Mitchell, "Clinical Lecture on Nervouess in the Male," 179, 180.
38. "Literary male hysterics" during this same period in the United States are discussed in Kristine Swenson's essay in this volume.
39. On the poet-novelist-physician as a cultural type, as well as art-science linkages generally in the late nineteenth century, see Mark S. Micale, "Introduction," in Micale, ed., *The Mind of Modernism: Medicine, Psychology, and the Cultural Arts in Europe and America, 1880–1940* (Stanford: Stanford University Press, 2004), 1–19; and Anne Stiles's introductory essay to this volume.
40. David M. Rein, *Weir Mitchell as a Psychiatric Novelist* (New York: International Universities Press, 1952). See also Earnest, *S. Weir Mitchell.*
41. Silas Weir Mitchell, "The Case of George Dedlow," *Atlantic Monthly* 18 (July 1866), 1–11.
42. Stanley Finger and Meredith Hustwit, "Five Early Accounts of Phantom Limb in Context: Paré, Descartes, Lemos, Bell, and Mitchell," *Neurosurgery* 52 (2003), 675–86. For the broader context, see Erin O'Connor's "'Fractions of Men': Engendering Amputation in Victorian Culture," *Comparative Studies in Society and History* 39 (1997), 742–77; and Long, *Rehabilitating Bodies*, 37–48.
43. See Long's excellent discussion of the novel in *Rehabilitating Bodies*, 49–57, although I see the relation between Mitchell's medical writings and literary output as less troubled and ambivalent than Long does.
44. Silas Weir Mitchell, *Roland Blake* (New York: Houghton, Mifflin, and Co., 1887), 345.
45. Mitchell, "Medical Department in the Civil War," 1449.
46. Edmund Wilson, *Patriotic Gore: Studies in the Literature of the American Civil War* (New York: Oxford University Press), ix.

Works Cited

The date of first publication (in serial or volume form) appears in parentheses after the title, where a later edition has been used.

Primary sources

Allen, Grant, *Physiological Aesthetics* (London: Henry King, 1877).
[Anon.], "The Charge Against Professor Ferrier," *The London Times* (18 November 1881), 10.
——, "Effects of Fright on the Mind," *Journal of Mental Science* 18 (July 1872), 234–5.
——, "Dr. Ferrier Accused," *The New York Times* (11 December 1881), 5.
——, "Dr. Ferrier's Researches," *The London Times* (19 November 1881), 12.
——, "Members and Associates: March, 1907," *Proceedings of the Society for Psychical Research* 19 (1907), 462–64.
——, *Professor Ferrier's Experiments on Monkeys' Brains* (London: Victoria Street Society, 1885).
——, review of Eugène Azam, "Double Consciousness with Periodic Loss of Memory (Amnesia)," *Mind* 1.3 (July 1876), 414–6.
——, review of J.R.L. Delboeuf, *Le magnetisme animal, Mind* 14 (1889), 470–1.
——, "The War," *The Lancet* (21 January 1871), 98–9.
Atherton, Gertrude, *Tower of Ivory*, 2 vols (Leipzig: Tauchnitz Verlag, 1910).
Bahr, Hermann, *Studien zur Kritik der Moderne* (Frankfurt: Rütten & Loening, 1894).
Bain, Alexander, *The Emotions and the Will* (1859), 3rd edn (London: Longmans, 1880).
——, *Mind and Body: The Theories of their Relation* (London: Henry King, 1873).
Baudelaire, Charles, *Richard Wagner in Paris* (Minden: Bruns, 1902).
Beard, George, *American Nervousness: Its Causes and Consequences* (1881; New York: Arno Press, 1972).
Bellamy, Edward, *Dr. Heidenhoff's Process* (1880; New York: AMS Press, 1969).
Benjamin, Walter, *Charles Baudelaire: A Lyric Poet in the Era of High Capitalism* (New York: Verso, 1997).
——, *Illuminations*, trans. Henry Zohn, ed. and intro. Hannah Arendt (New York: Schoken Books, 1988).
Blumer, G. Alder, "A Case of Perverted Sexual Instinct," *American Journal of Insanity* 39 (July 1882), 22–35.
Boggs, Alex, "Paris after the Double Siege," *The Lancet* (8 July 1871), 75.
Breton, André, "Le Message automatique," *Minotaure* 3–4 (December 1933), 41–64.
——, *Œuvres complètes*, 2 vols (Paris: Gallimard, 1992).
Breuer, Josef and Sigmund Freud, *Studies on Hysteria* (1895), trans. and ed. James Strachey and Anna Freud (New York: Basic Books, 1957).
Brontë, Charlotte, *Jane Eyre* (1847), ed. Margaret Smith and Sally Shuttleworth (Oxford: Oxford University Press, 2000).

Burr, George, "Cases of Injuries of the Nervous Centres, from Explosion of Shells, without Wound or Contusion," *New York Medical Journal* 1 (September 1865), 428–32.
Carpenter, William Benjamin, "On the Doctrine of Human Automatism: Part II," *Contemporary Review* 25 (1875), 940–62.
——, *Principles of Mental Physiology* (New York: Appleton, 1874).
Charcot, Jean-Martin and Valentin Magnan, "Inversion du sens génital," *Archives de Neurologie* 3 (1882), 53–60 and 4 (1882), 296–322.
Cheyne, George, *The English Malady* (London: G. Strahan, 1733).
Cobbe, Frances Power, "Vivisection and Its Two-Faced Advocates," *The Contemporary Review* 41 (1882), 610–26.
Collins, Wilkie, *Heart and Science: A Story of the Present Time*, 2 vols (Leipzig: Bernard Tauchnitz, 1883).
——, *The Law and the Lady* (1875), ed. Jenny Bourne Taylor (Oxford: Oxford World's Classics, 1992).
Corelli, Marie, *Ardath: The Story of a Dead Self* (1889; London: Methuen, 1907).
——, *Wormwood: A Drama of Paris* (Chicago: M.A. Donohue, 1890).
Da Costa, Jacob Mendez, "On Irritable Heart: A Clinical Study of a Form of Functional Cardiac Disorder and Its Consequences," *American Journal of the Medical Sciences* 121 (January 1871), 17–52.
[Dallas, Eneas Sweetland], "The First Principle of Physiognomy," *Cornhill* 4 (1861), 569–81.
[——], "On Physiognomy," *Cornhill* 4 (1861), 472–81.
Darwin, Charles, *Darwin*, ed. Philip Appleman (New York: Norton, 1979).
Dickens, Charles, *Bleak House* (1852–53), ed. Andrew Sanders (London: J.M. Dent, 1994).
Dostoyevsky, Fyodor, *Notes from the Underground* (1864) and *The Double* (1846), ed. Jessie Coulson (London: Penguin Books, 2003).
Du Maurier, George, *Trilby* (1894; Oxford: Oxford University Press, 1995).
Eliot, George, *Daniel Deronda* (1876), ed. Graham Handley (Oxford: Oxford World's Classics, 1984).
——, *Middlemarch* (1871–2), ed. David Carroll (Oxford: Oxford World's Classics, 1998).
Esquirol, Jean Étienne Dominique, *Mental Maladies: A Treatise on Insanity*, trans. E.K. Hunt (Philadelphia: Lea and Blanchard, 1845).
Ferrier, David, "Experimental Researches in Cerebral Physiology and Pathology," *West Riding Lunatic Asylum Medical Reports* 3 (1873), 30–96.
——, *The Functions of the Brain* (1876; London: Dawson's, 1966).
——, *The Localization of Cerebral Disease* (London: Smith, Elder, 1878).
——, "Pathological Illustrations of Brain Function," *West Riding Lunatic Asylum Medical Reports* 4 (1874), 30–62.
Flitch, J.E. Crawford, *Modern Dancing and Dancers* (Philadelphia: J.B. Lippincott, 1912).
Flournoy, Theodore, "Chorégraphie somnambulique. Le cas de Magdeleine G.," *Archives de psychologie* 3 (July 1904), 357–74.
Freud, Sigmund, *The Freud Reader*, ed. Peter Gay (New York and London: W.W. Norton, 1989).
——, *The Standard Edition of the Complete Psychological Works of Sigmund Freud*, ed. James Strachey and Anna Freud, 24 vols (London: Hogarth Press, 1981).

Fritsch, Gustav and Eduard Hitzig, "Über die elektrische Erregbarkeit des Grosshirns," *Archiv für Anatomie, Physiologie und wissenschaftliche Medicin* 37 (1870), 300–32.
Fuchs, Hanns, *Richard Wagner und die Homosexualität* (Berlin: Barsdorf, 1903).
Furness, Ray, ed., *The Dedalus Book of German Decadence: Voices of the Abyss* (New York: Hippocrene, 1994).
Furst, Lilian R., ed., *Medical Progress and Social Reality: A Reader in Nineteenth-Century Medicine and Literature* (Albany: State University of New York Press, 2000).
Galton, Francis, "Hereditary Talent and Character: Part I," *Macmillan's Magazine* 12 (1865), 157–66.
Galvani, Luigi, "De viribus electricitatis in motu musculari. Commentarius," *De Bonoiensi Scientiarum et Artium Instituo atque Academie Commentari* (1791), 7: 363–418.
Gibson, James, *Memoirs of the Brave: A Brief Account of the Battles of the Alma, Balaklava, and Inkerman, with Biographies of the Killed and a List of the Wounded* (1889; London: The London Stamp Exchange, 1890).
Gissing, George, *The Whirlpool* (London: Lawrence and Bullen, 1897).
Gourmont, Rémy de, *La Culture des idées* (Paris: Mercure de France, 1900).
Grand, Sarah, "The Man of the Moment," *North American Review* 158 (1894), 620–628.
——, "The New Aspect of the Woman Question," *North American Review* 158 (1894), 270–7.
Green, Malcolm, ed., *The Golden Bomb: Phantastic German Expressionist Stories* (Edinburgh: Polygon, 1993).
Hammond, William M., *Sexual Impotence in the Male* (New York: Birmingham, 1883).
Hartmann, Eduard von, *Philosophie de l'inconscient*, trans. D. Nolen (1869; Paris: Ballière, 1877).
Hering, Ewald, "Memory as a General Function of Organized Matter," *The Open Court, a Quarterly Magazine* 1.6 (April 1887), 141–7.
Holmes, Oliver Wendell, *A Mortal Antipathy*, 3rd edn (Boston: Houghton, Mifflin, 1886).
Howe, Mark de Wolfe, ed., *Touched with Fire: Civil War Letters of Oliver Wendell Holmes, Jr., 1861–1864* (Cambridge: Harvard University Press, 1946).
Howells, William Dean, *Dr. Breen's Practice, Atlantic Monthly* 48 and 49 (1881–82).
Huneker, James Gibbons, *Melomaniacs* (1902; New York: Charles Scribner's Sons, 1920).
——, *Overtures* (London: Ibister and Co., 1904).
Huysmanns, Joris-Karl, *Against Nature* (1884), trans. Robert Baldick (New York: Penguin, 1959).
Ireland, Mrs. Alexander, ed., *Selections from the Letters of Geraldine Endsor Jewsbury to Jane Welsh Carlyle* (London: Longmans, Green and Co., 1892).
James, William, *Psychology* (1892; New York: Henry Holt, 1908).
Janet, Pierre, *L'automatisme psychologique* (Paris: Alcan, 1889).
Jentsch, Ernst, *Musik und Nerven* (Wiesbaden: Bergmann, 1911).
Jordan, John Furneaux, "On Shock After Surgical Operations and Injuries," *Special Enquiries; Including the Hastings Essay on Shock, the Treatment of Surgical Inflammations, and Numerous Clinical Lectures*, 2nd edn (London: J.A. Churchill, 1881), 1–60.

Keen, William W., Silas Weir Mitchell, and George R. Morehouse, "On Malingering, especially in Regard to Simulation of Diseases of the Nervous System," *American Journal of Medical Science* 48 (1864), 367–94.

Kent, James Tyler, *Lectures on Homeopathic Philosophy* (1900; Richmond, California: North Atlantic Books, 1979).

Kiernan, James G., "Insanity. Lecture XXVI.—Sexual Perversion," *Detroit Lancet* 7.11 (May 1884), 481–4.

——, "Perverted Sexual Instinct," *Chicago Medical Journal and Examiner* 48 (1884), 263–5.

Kinglake, Alexander, *The Invasion of Crimea: Its Origin, and an Account of its Progress Down to the Death of Lord Raglan*, 6 vols (New York: Harper & Brothers, 1881).

Krafft-Ebing, Richard Freiherr von, *Neue Forschungen auf dem Gebiete des Psychopathia Sexualis* (Stuttgart: Ferdinand Enke, 1890).

——, *Über gesunde und kranke Nerven* (Tübingen: Verlag der Laupp'schen Buchhandlung, 1885).

Lake, Colonel Atwell C.B., *Kars and Our Captivity in Russia* (London: Richard Bentley, 1856).

Leroux, Gaston, *The Phantom of the Opera* (London: Penguin Popular Classics, 1995).

Luddy, Maria, ed., *The Crimean Journals of the Sisters of Mercy, 1854–56* (Dublin: Four Courts Press, 2004).

Lugné-Poë, Aurélien, *La Parade sous les étoiles. Souvenirs de theâtre, 1902–1912* (Paris: Gallimard, 1933).

Maeterlinck, Maurice, *The Buried Temple*, trans. Alfred Sutro (London: Ruskin House, 1902).

——, "La Chance," *Revue de Paris* (15 March 1901), 241–66.

Magnin, Émile, *Art et l'hypnose. Interpretation plastique d'oeuvres littéraires et musicales* (Paris: Félix Alcan, 1907).

——, "Catalepsie et extase musicale," *Revue spirite* (July 1903), 3–10.

Maher, Michael, *Psychology Empirical and Rational*, 8th edn (London: Longmans, Green, 1915).

Maier, Friedrich, "Die 'Schlaftänzerin' Madeleine G. in München," *Psychische Studien* 31.5 (1904), 239–46.

Mann, Thomas, *Buddenbrooks*, trans. H.T. Lowe-Porter (New York: Alfred A. Knopf, 1924).

——, *Stories of Three Decades*, trans. H.T. Lowe-Porter (London: Secker and Warburg, 1946).

Marx, Henry, "Madeleine: Two Reviews," *The Drama Review* 22.2 (June 1978), 29–31.

Maudsley, Henry, *Body and Mind: An Inquiry Into Their Connection And Mutual Influence, Specially In Reference To Mental Disorders*, revised edn (New York: Appleton, 1885).

Mitchell, Silas Weir, *The Autobiography of a Quack* (1867) and *The Case of George Dedlow* (1866) (New York: The Century, 1900).

——, "The Case of George Dedlow," *Atlantic Monthly* 18 (July 1866), 1–11.

——, "Clinical Lecture on Nervousness in the Male," *The Medical News and Library* 35 (December 1877), 177–84.

——, *Injuries of Nerves and Their Consequences* (1872; New York: Dover, 1965).

——, "The Medical Department in the Civil War," *Journal of the American Medical Association* 62 (May 1914), 1445–50.

——, "Phantom Limbs," *Lippincott's Magazine of Popular Literature and Science* 8 (1871), 563–9.

——, *Roland Blake* (New York: Houghton, Mifflin, and Co., 1887).
——, George R. Morehouse, and William W. Keen, *Gunshot Wounds and Other Injuries of Nerves* (Philadelphia: J.B. Lippincott & Co., 1864).
Moll, Albert, *Hypnotism, Including a Study of the Chief Points of Psycho-Therapeutics and Occultism*, trans. Arthur F. Hopkirk (London: Walter Scott, 1909).
——, *Psychologie und Charakterologie der Okkultisten* (Stuttgart: Ferdinand Enke, 1929).
Moore, George, *Evelyn Innes* (New York: Appleton and Co., 1898).
Moreau, Jacques, *La psychologie morbide dans ses rapports avec la philosophie de l'histoire ou de l'influence des névropathies sur le dynamisme intellectuel* (Paris: Librairie Victor Masson, 1859).
Morris, Edwin, *A Practical Treatise on Shock* (1867; Philadelphia: J.B. Lippincott, 1868).
Morselli, Enrico, "Dysmorphophobia and Taphephobia: Two Hitherto Undescribed forms of Insanity with Fixed Ideas," ed. Laurence Jerome, *History of Psychiatry* 12 (2001), 103–14.
Myers, Frederic W. H., "Multiplex Personality," *The Ninteenth Century* 20 (November 1886), 648–66.
Nelson, Carolyn Christensen, ed., *A New Woman Reader: Fiction, Articles, and Drama of the 1890s* (Peterborough, Ontario: Broadview Press, 2001).
Nietzsche, Friedrich, *The Birth of Tragedy* (1872) *and The Case of Wagner* (1888), trans. Walter Kauffmann (New York: Vintage, 1967).
Nordau, Max, *Degeneration* (1892), trans. George L. Mosse (Lincoln: University of Nebraska Press, 1993).
Otis, Laura, ed., *Literature and Science in the Nineteenth Century. An Anthology* (New York: Oxford University Press, 2002).
Page, Herbert W., *Injuries of the Spine and Spinal Cord Without Apparent Lesion, and Nervous Shock, in their Surgical and Medico-Legal Aspects* (London: J. & A. Churchill, 1883).
——, *Railway Injuries, With Special Reference to Those of the Back and Nervous System, in Their Medico-Legal and Clinical Aspects* (London: Griffin, 1891).
Panizza, Oskar, "Bayreuth und Homosexualität: eine Erwägung," *Die Gesellschaft: Monatsschrift für Literatur, Kunst und Socialpolitik* 11 (1895), 88–92.
Phelps, Elizabeth Stuart, *Doctor Zay* (1882), ed. Michael Sartisky (New York: The Feminist Press, 1987).
——, "Zerviah Hope," *Scribner's Monthly* 21.1 (November 1880), 78–89.
Plato, *The Works of Plato*, ed. Thomas Taylor, 5 vols (Frome: The Prometheus Trust, 1996).
Proctor, Richard, "Dual Consciousness," *The Cornhill Magazine* 35 (January 1877), 86–105.
Proust, Adrien, *Treatment of Neurasthenia*, trans. Peter Campbell Smith (London: Henry Kimpton, 1902).
Puschmann, Theodor, *Richard Wagner: eine psychiatrische Studie* (Berlin: Behr's, 1872).
Putnam, James J., "The Medico-Legal Significance of Hemianaesthesia After Concussion Accidents," *American Journal of Neurology and Psychiatry* 3 (1884), 507–16.
——, "Recent Investigations into the Pathology of So-called Concussion of the Spine," *Boston Medical and Surgical Journal* 109 (1883), 217–20.
Pye-Smith, P.H., "Address," *Nature* (28 August 1879), 408–13.

Reynolds, Russell, "Remarks on Paralysis and Other Disorders of Motion and Sensation, Dependent on Idea," *British Medical Journal* 2 (1869), 483–5.
Ribot, Théodule, *The Diseases of Memory* (1881; New York: J. Fitzgerald, 1883).
Roberts, Franz, *Die Schlaftänzerin Madeleine G.; ein Protest gegen den Missbrauch der Wissenschaft* (Munich: n.p., 1904).
"R.V.," "Un article sur les Décadents," *Le Décadent. Revue littéraire bi-mensuelle* 3.25 (15–31 December 1888), 13–16.
Schreiner, Olive, *Woman and Labour* (1911; London: Virago Press, 1978).
Schrenck-Notzing, Albert von, "Albert von Keller als Malerpsychologe und Metapsychiker," *Psychische Studien* 48 (1921), 201–10.
—— and Otto Schultze, *Die Traumtänzerin Magdeleine G.: Eine psychologische Studie über Hypnose und dramatische Kunst* (Stuttgart: F. Encke, 1904).
Shaw, J.C. and G.N. Ferris, "Perverted Sexual Instinct," *Journal of Nervous and Mental Disease* 10.2 (April 1883), 185–204.
Sherman, William Tecumseh, Address before graduating class, Michigan Military Academy, 19 June 1879.
Spencer, Herbert, *The Principles of Psychology* (1855), 2nd edn, 2 vols (London: Williams and Norgate, 1870–72).
Spitzka, E.C., "A Historical Case of Sexual Perversion," *Chicago Medical Review* 4.4 (August 1881), 378–9.
Spitzer, Daniel, *Richard Wagners Briefe an eine Putzmacherin* (Vienna: Carl Konegen, 1906).
——, *Verliebte Wagnerianer* (Wien: Julius Klinkhardt, 1880).
Stevenson, Robert Louis, *Strange Case of Dr. Jekyll and Mr. Hyde* (1886), ed. Katherine Linehan (New York: Norton, 2003).
Stoker, Bram, *The Snake's Pass* (1890; Dingle, Ireland: Brandon, 1990).
Symons, Arthur, *Eleonora Duse* (London: Ayer, 1927).
Taylor, Jenny Bourne and Sally Shuttleworth, eds, *Embodied Selves: An Anthology of Psychological Texts, 1830–1890* (New York: Oxford University Press, 1998).
Tolstoy, Leo, *The Kreutzer Sonata* (London: Walter Scott, 1898).
Tuke, Daniel Hack, *Illustrations of the Influence of the Mind upon the Body in Health and Disease: Designed to Elucidate the Action of the Imagination*, 2nd edn, (1872; Philadelphia: Henry C. Lea's Son & Co., 1884).
——, ed., *A Dictionary of Psychological Medicine* (Philadelphia: P. Blackiston, 1892).
Tuttle, Jennifer, ed., "Letters from Elizabeth Stuart Phelps (Ward) to S. Weir Mitchell, M.D., 1884–1897," *Legacy: Journal of American Women Writers* 17.1 (2000), 83–94.
Verlaine, Paul, *One Hundred and One Poems by Paul Verlaine: A Bilingual Edition*, trans. Norman R. Shapiro (Chicago: University of Chicago Press, 1999).
Walton, G.L., "Case of Hysterical Hemianaesthesia, Convulsions and Motor Paralysis Brought on by a Fall," *Boston Medical and Surgical Journal* 111 (1884), 558–9.
——, "Case of Typical Hysterical Hemianesthesia in a Man Following Injury," *Archives of Medicine* 10 (1883), 88–95.
Ward, Andrew, *Our Bones are Scattered: The Cawnpore Massacre and the Indian Mutiny of 1857* (New York: Henry Holt, 1996).
Welch, Spencer Glasgow, *A Confederate Surgeon's Letters to His Wife* (New York: Neale Publishing, 1911).
Wells, H.G., *Experiment in Autobiography* (New York: Macmillan, 1934).
——, *The Invisible Man* (1897), ed. Patrick Parrinder (London: Penguin, 2005).

——, *The Island of Dr. Moreau* (1896): *A Variorum Text,* ed. Robert M. Philmus (Athens, Georgia: University of Georgia Press, 1993).

Wilde, Oscar, *The Picture of Dorian Gray* (1890, revised 1891), ed. Robert Mighall (London: Penguin, 2000).

Secondary sources

Amacher, Peter, "Freud's Neurological Education and its Influence on Psychoanalytic Theory," *Psychological Issues* 4.4 (1965), 21–8.

[Anon.], *Franz von Stuck und der Tanz. Der Tanz um die Jahrhundertwende* (Tettenweis: Förderkreis Franz von Stuck Geburtshaus Tettenweis e.V., 2001).

Asendorf, Christoph, *Batteries of Life: On the History of Things and Their Reception in Modernity* (1984), trans. Don Reneau (Berkeley: University of California Press, 1993).

Beebe, G.W. and M.E. DeBakey, *Battle Casualties: Incidence, Mortality, and Logistic Considerations* (Springfield: Charles C. Thomas, 1952).

Benson, Herbert, *The Relaxation Response* (1975; New York: HarperCollins, 2001).

Bennett, Mary Angela, *Elizabeth Stuart Phelps* (Philadelphia: University of Pennsylvania Press, 1939).

Bergonzi, Bernard, *The Early H.G. Wells* (Manchester: Manchester University Press, 1961).

Bogdan, Robert, *Freak Show: Presenting Human Oddities for Amusement and Profit* (Chicago: University of Chicago Press, 1988).

Bowen, Roger, "Science, Myth, and Fiction in H.G. Wells's *Island of Dr. Moreau,*" *Studies in the Novel* 8 (1976), 318–35.

Brunner, José, "Identifications, Suspicions, and the History of Traumatic Disorders," *Harvard Review of Psychiatry* 10.3 (2002), 179–84.

Brunnert, Klaus, *Nostalgie in der Geschichte der Medizin* (Düsseldorf: Triltsch, 1984).

Cantor, Geoffrey and Sally Shuttleworth, eds, *Science Serialized: Representations of the Sciences in Nineteenth-Century Periodicals* (Cambridge: The MIT Press, 2004).

Caplan, Eric, *Mind Games: American Culture and the Birth of Psychotherapy* (Berkeley: University of California Press, 1998).

Cervetti, Nancy, "S. Weir Mitchell Representing 'a hell of pain:' From Civil War to Rest Cure," *Arizona Quarterly* 59 (2003), 69–96.

Clapp-Itnyre, Alisa, *Angelic Airs, Subversive Songs: Music as Social Discourse in the Victorian Novel* (Athens: Ohio University Press, 2002).

Clarke, Edwin and L.S. Jacyna, *Nineteenth-Century Origins of Neuroscientific Concepts* (Berkeley: University of California Press, 1987).

Cusick, Suzanne, "Gendering Modern Music: Thoughts on the Monteverdi-Artusi Controversy," *Journal of the American Musicological Association* 46 (Spring 1993), 1–25.

Dames, Nicholas, *Amnesiac Selves: Nostalgia, Forgetting and British Fiction 1810–1870* (Oxford: Oxford University Press, 2001).

Danziger, Kurt, *Naming the Mind: How Psychology Found its Language* (London: Sage, 1997).

Dardenne, Émilie, "Frances Power Cobbe (1822–1904) et la naissance du mouvement contre l'éxperimentation animale," *Cahiers victoriens et edouardiens* 59 (2004), 171–80.

Davis, Michael, *George Eliot and Nineteenth-Century Psychology: Exploring the Unmapped Country* (Aldershot: Ashgate, 2006).
Dean, Eric T. Jr., *Shook Over Hell: Post-Traumatic Stress, Vietnam, and the Civil War* (Cambridge: Harvard University Press, 1997).
——, "We Will All Be Lost and Destroyed: Post-Traumatic Stress Disorder and the Civil War," *Civil War History* 37 (June 1991), 138–53.
D'Emilio, John and Estelle B. Freedman, *Intimate Matters: A History of Sexuality in America* (New York: Harper and Row, 1988).
Dixon, Thomas, *From Passions to Emotions: The Creation of a Secular Psychological Category* (Cambridge: Cambridge University Press, 2003).
Dowbiggin, Ian, "Back to the Future: Valentin Magnan, French Psychiatry, and the Classification of Mental Diseases, 1885–1925," *Social History of Medicine* 9.3 (1996), 383–409.
Drinka, George Frederick, *The Birth of Neurosis: Myth, Malady, and the Victorians* (New York: Simon and Schuster, 1984).
Earnest, Ernest, *S. Weir Mitchell: Novelist and Physician* (Philadelphia: University of Pennsylvania Press, 1950).
Eickhorst, William, *Decadence in German Fiction* (Denver: Alan Swallow, 1953).
Ellenberger, Henri F., *The Discovery of the Unconscious: The History and Evolution of Dynamic Psychiatry* (New York: Basic Books, 1970).
Finger, Stanley, *Origins of Neuroscience: A History of Explorations into Brain Function* (New York: Oxford University Press, 1994).
—— and Meredith Hustwit, "Five Early Accounts of Phantom Limb in Context: Paré, Descartes, Lemos, Bell, and Mitchell," *Neurosurgery* 52 (2003), 675–86.
Foucault, Michel, *The History of Sexuality*, vol. 1: *The Will to Knowledge* (1976), trans. Robert Hurley (New York: Random House, 1980).
Frank, Lawrence, *Victorian Detective Fiction and the Nature of Evidence: The Scientific Investigations of Poe, Dickens, and Doyle* (New York: Palgrave Macmillan, 2003).
French, Richard, *Antivivisection and Medical Science in Victorian Society* (Princeton: Princeton University Press, 1975).
Frost, Linda, *Never One Nation: Freaks, Savages, and Whiteness in US Popular Culture, 1850–1877* (Minneapolis: University of Minnesota Press, 2005).
Fuchs, Georg, *Sturm und Drang in München um die Jahrhundertwende* (Munich: Verlag Georg D.W. Callwey, 1936).
Fulcher, Jane, "Melody and Morality: Rousseau's Influence on French Music Criticism," *International Review of the Aesthetics and Sociology of Music* 11.1 (June 1980), 45–57.
Fuller, Sophie, *The Idea of Music in Victorian Fiction* (Aldershot: Ashgate, 2004).
Furst, Lilian R., *Between Doctors and Patients: The Changing Balance of Power* (Charlottesville: University of Virginia Press, 1998).
——, ed., *Women Healers and Physicians: Climbing a Long Hill* (Lexington: University Press of Kentucky, 1997).
Gabriel, Richard, *No More Heroes: Madness & Psychiatry in War* (New York: Hill and Wang, 1987).
Gallini, Clara, *La sonnambula meravigliosa. Magnetismo e ipnotismo nell'Ottocento italiano* (Milan: Feltrinelli, 1983).
Gijswijt-Hofstra, Marijke and Roy Porter, eds, *Cultures of Neurasthenia from Beard to the First World War* (New York: Rodopi, 2001).

Gillispie, Charles Coulston, ed., *Dictionary of Scientific Biography*, 14 vols (New York: Charles Scribner's Sons, 1970–80).

Gilman, Sander L., *Creating Beauty to Cure the Soul: Race and Psychology in the Shaping of Aesthetic Surgery* (Durham: Duke University Press, 1998).

——, *Jewish Frontiers: Essays on Bodies, Histories and Identities* (New York: Palgrave Macmillan, 2003).

——, *Making the Body Beautiful: A Cultural History of Aesthetic Surgery* (Princeton: Princeton University Press, 1999).

——, "Strauss, the Pervert, and Avant-Garde Opera of the *Fin de Siècle*," *New German Critique* 43 (Winter 1998), 35–68.

Goetz, Christopher G., Michel Bonduelle, and Toby Gelfand, *Charcot: Constructing Neurology* (New York: Oxford University Press, 1995).

Goldstein, Jan, *Console and Classify: The French Psychiatric Profession in the Nineteenth Century* (New York: Cambridge University Press, 1987).

Gosling, F. G., *Before Freud: Neurasthenia and the American Medical Community, 1870–1910* (Urbana and Chicago: University of Illinois Press, 1987).

Gouk, Penelope, ed., *Musical Healing in Cultural Contexts* (Aldershot: Ashgate, 2000).

Hacking, Ian, *Rewriting the Soul: Multiple Personality and the Sciences of Memory* (Princeton: Princeton University Press, 1995).

Hagner, Michael, ed., *Ecce Cortex: Beiträge zur Geschichte des modernen Gehirns.* (Göttingen: Wallstein, 1999).

Hall, J., G. Zilboorg, and H.A. Bunder, eds, *One Hundred Years of American Psychiatry* (New York: Columbia University Press, 1944).

Haller, John S., *American Medicine in Transition, 1840–1910* (Urbana: University of Illinois Press, 1981).

Harrington, Anne, *Medicine, Mind, and the Double Brain: A Study in Nineteenth-Century Thought* (Princeton: Princeton University Press, 1987).

Hartley, Lucy, *Physiognomy and the Meaning of Expression in Nineteenth-Century Culture* (Cambridge: Cambridge University Press, 2001).

Haynes, Roslynn D., "The Unholy Alliance of Science in *The Island of Dr. Moreau*," *The Wellesian* 11 (1988), 13–24.

——, "Wells's Debt to Huxley and the Myth of Dr. Moreau," *Cahiers victoriens et édouardiens* 13 (1981), 31–41.

Hennelly, Mark M., "Reader Vivisection in *The Island of Dr. Moreau*," *Essays in Arts and Sciences* 9 (1980), 217–33.

Horden, Peregrine, ed., *Music as Medicine: The History of Music Therapy Since Antiquity* (Aldershot: Ashgate, 2000).

Hughes, Robert, *The Shock of the New* (New York: Knopf, 1981).

Hurley, Kelly, *The Gothic Body: Sexuality, Materialism, and Degeneration at the* Fin de Siècle (New York: Cambridge University Press, 1996).

Jelavich, Peter, *Munich and Theatrical Modernism: Politics, Playwriting, and Performance, 1890–1914* (Cambridge: Harvard University Press, 1985).

John, Eckhard, *Musikbolschewismus: Die Politisierung der Musik in Deutschland 1918–38* (Stuttgart: Metzler Verlag, 1994).

Jones, Edgar and Simon Wessely, "Case of Chronic Fatigue Syndrome After Crimean War and Indian Mutiny," *British Medical Journal* 319 (1999), 1645–1647.

——, "Psychiatric Battle Casualties: an Intra- and Interwar Comparison," *The British Journal of Psychiatry* 178 (2001), 242–47.

——, *Shell-Shock to PTSD: Military Psychiatry from 1900 to the Gulf War* (Hove: Psychology Press, 2006).
Kaplan, E. Ann, *Trauma Culture: The Politics of Terror and Loss in Media and Literature* (New Brunswick: Rutgers University Press, 2005).
Katz, Jonathan Ned, *Gay/Lesbian Almanac: A New Documentary* (New York: Harper and Row, 1983).
Kaufmann, Martin, *Homeopathy in America; the Rise and Fall of a Medical Heresy* (Baltimore: Johns Hopkins University Press, 1971).
Keller, Thomas, "Railway Spine Revisited: Traumatic Neurosis or Neurotrama?," *Journal of the History of Medicine and Allied Sciences* 50 (October 1995), 507–24.
Kirschmann, Anne, *A Vital Force: Women in American Homeopathy* (New Brunswick: Rutgers University Press, 2004).
Koppen, Erwin, *Dekadenter Wagnerismus* (Berlin: Walter de Gruyter, 1973).
Krobb, Florian, "'*Die Kunst der Vater todtet das Leben der Enkel*': Decadence and Crisis in *Fin-de-Siècle* German and Austrian Discourse," *New Literary History* 35.4 (2004), 547–62.
LaCapra, Dominick, *Madame Bovary on Trial* (Ithaca: Cornell University Press, 1982).
Lansbury, Coral, "Gynaecology, Pornography, and the Antivivisection Movement," *Victorian Studies* 28 (1985), 413–37.
——, *The Old Brown Dog: Women, Workers and Vivisection in Edwardian England* (Madison: University of Wisconsin Press, 1985).
Layton, Susan, *Russian Literature and Empire: Conquest of the Caucasus from Pushkin to Tolstoy* (Cambridge: Cambridge University Press, 1994).
Le Gallienne, Eva, *The Mystic in the Theatre: Eleonora Duse* (Carbondale: Southern Illinois University Press, 1973).
Leavis, F.R., *Two Cultures? The Significance of C.P. Snow* (London: Chatto and Windus, 1962).
LeDoux, Joseph, *The Emotional Brain, the Mysterious Underpinnings of Emotional Life* (New York: Simon and Schuster, 1996).
Leys, Ruth, *Trauma, A Genealogy* (Chicago: University of Chicago Press, 2000).
Lippman, Edward A., "The Sources and Development of the Ethical View of Music in Ancient Greece," *The Musical Quarterly* 49.2 (April 1963), 188–209.
Long, Lisa A., *Rehabilitating Bodies: Health, History, and the American Civil War* (Philadelphia: University of Pennsylvania Press, 2004).
Lutz, Tom, *American Nervousness, 1903: An Anecdotal History* (Ithaca: Cornell University Press, 1991).
Macmillan, Malcolm, *Freud Evaluated: The Completed Arc* (Cambridge, Massachusetts: MIT Press, 1997).
Masters, Brian, *Now Barrabas was a Rotter: The Extraordinary Life of Marie Corelli* (London: Hamish Hamilton, 1978).
Matus, Jill L., "Trauma, Memory, and Railway Disaster: The Dickensian Connection," *Victorian Studies* 43.3 (2001), 413–36.
McClary, Susan, *Feminine Endings: Music, Gender and Sexuality* (Minneapolis: University of Minnesota Press, 1991).
McClatchie, Stephen, *Analyzing Wagner's Operas: Alfred Lorenz and German Nationalist Ideology* (Rochester: University of Rochester Press, 1998).
McGuinness, Patrick, ed., *Symbolism, Decadence and the* Fin-de-Siècle (Exeter: University of Exeter Press, 2000).

McLaren, Angus, *The Trials of Masculinity: Policing Sexual Boundaries, 1870–1930* (Chicago: University of Chicago Press, 1997).
Méheust, Bertrand, *Somnambulisme et médiumnité, 1784–1930*, 2 vols (Le Plessis: Éditions Synthélabo, 1991).
Mendelson, Danuta, *The Interfaces of Medicine and Law: The History of the Liability for Negligently Caused Psychiatric Injury (Nervous Shock)* (Aldershot: Ashgate, 1998).
Menke, Richard, "Fiction as Vivisection: G.H. Lewes and George Eliot," *ELH* 67.2 (2000), 617–53.
Micale, Mark S., *Approaching Hysteria: Disease and Its Interpretations* (Princeton: Princeton University Press, 1995).
——, "Charcot and the Idea of Hysteria in the Male: Gender, Mental Science, and Medical Diagnosis in Late Nineteenth-Century France," *Medical History* 34 (1990), 363–411.
——, ed., *The Mind of Modernism: Medicine, Psychology, and the Cultural Arts in Europe and America, 1880–1940* (Stanford: Stanford University Press, 2004).
—— and Paul Lerner, eds, *Traumatic Pasts: History, Psychiatry, and Trauma in the Modern Age, 1870–1930* (New York: Cambridge University Press, 2001).
—— and Roy Porter, eds, *Discovering the History of Psychiatry* (New York: Oxford University Press, 1994).
Mighall, Robert, *A Geography of Victorian Gothic Fiction* (New York: Oxford University Press, 1999).
Miller, Neil, *Out of the Past: Gay and Lesbian History from 1869 to the Present* (New York: Vintage, 1995).
Mondimore, Francis Mark, *A Natural History of Homosexuality* (Baltimore: Johns Hopkins University Press, 1996).
Morantz-Sanchez, Regina, *Sympathy and Science. Women Physicians in American Medicine* (New York: Oxford University Press, 1985).
Nussbaum, Martha N., *Upheavals of Thought: The Intelligence of Emotions* (Cambridge: Cambridge University Press, 2001).
Oatley, Keith, *A Brief History of Emotions* (Oxford: Basil Blackwell, 2004).
Ochs, Sidney, *A History of Nerve Functions: From Animal Spirits to Molecular Mechanisms* (Cambridge: Cambridge University Press, 2004).
O'Connor, Erin, "'Fractions of Men': Engendering Amputation in Victorian Culture," *Comparative Studies in Society and History* 39 (1997), 742–77.
Oosterhuis, Harry, *Stepchildren of Nature: Krafft-Ebing, Psychiatry, and the Making of Sexual Identity* (Chicago: University of Chicago Press, 2000).
Oppenheim, Janet, *"Shattered Nerves": Doctors, Patients, and Depression in Victorian England* (New York: Oxford University Press, 1991).
Otis, Laura, *Membranes: Metaphors of Invasion in Nineteenth-Century Literature, Science and Politics* (Baltimore: Johns Hopkins University Press, 1999).
——, *Networking: Communicating with Bodies and Machines in the Nineteenth Century* (Ann Arbor: University of Michigan Press, 2001).
Painter, Karen, "The Sensuality of Timbre: Responses to Mahler and Modernity at the *Fin de Siècle*," *Nineteenth Century Music* 18.3 (Spring 1995), 236–56.
Panksepp, Jaak, *Affective Neuroscience: The Foundations of Human and Animal Emotions* (Oxford: Oxford University Press, 1998).
Payne, David, *The Reenchantment of Nineteenth-Century Fiction. Dickens, Thackeray, George Eliot and Serialization* (New York: Palgrave Macmillan, 2005).

Pedlar, Valerie, "Experimentation or Exploitation? The Investigations of David Ferrier, Dr. Benjulia, and Dr. Seward," *Interdisciplinary Science Reviews* 28 (2003), 169–74.

Peiss, Kathy and Christina Simmons, eds, *Passion and Power: Sexuality in History* (Philadelphia: Temple University Press, 1989).

Phillips, Katharine A., *The Broken Mirror: Understanding and Treating Body Dysmorphic Disorder* (1986; Oxford: Oxford University Press, 1998).

Pick, Daniel, *Svengali's Web: The Alien Enchanter in Modern Culture* (New Haven: Yale University Press, 2000).

Pierrot, Jean, *The Decadent Imagination, 1880–1900*, trans. Derek Colman (Chicago: University of Chicago Press, 1981).

Plummer, Kenneth, ed., *The Making of the Modern Homosexual* (Totowa, New Jersey: Barnes and Noble, 1981).

Rein, David M., *Weir Mitchell as a Psychiatric Novelist* (New York: International Universities Press, 1952).

Richardson, Alan, *British Romanticism and the Science of the Mind* (New York: Cambridge University Press, 2001).

—— and Ellen Spolsky, eds, *The Work of Fiction: Cognition, Culture and Complexity* (Burlington, Vermont: Ashgate, 2004).

Richardson, Angelique, *Love and Eugenics in the Late Nineteenth Century. Rational Reproduction and the New Woman* (New York: Oxford University Press, 2003).

—— and Chris Willis, eds, *The New Woman in Fiction and in Fact* (New York: Palgrave Macmillan, 2001).

Rosario, Vernon A., *The Erotic Imagination: French Histories of Perversity* (New York: Oxford University Press, 1997).

Rosen, George, "Nostalgia: A Forgotten Psychological Disorder," *Clio Medica* 10 (1975), 28–51.

Rosenberg, Edgar, *From Shylock to Svengali* (Stanford: Stanford University Press, 1960).

Roth, Michael, "Hysterical Remembering," *Modernism/Modernity* 3.2 (1996), 1–30.

Rotundo, E. Anthony, *American Manhood: Transformations in Masculinity from the Revolution to the Modern Era* (New York: Basic Books, 1993).

Rupke, Nicolaas A., ed., *Vivisection in Historical Perspective* (New York: Routledge, 1990).

Rylance, Rick, *Victorian Psychology and British Culture, 1850–1880* (New York: Oxford University Press, 2000).

Sacks, Oliver, *A Leg to Stand On* (1984; New York: Touchstone, 1998).

——, *The Man Who Mistook His Wife for a Hat and Other Clinical Tales* (1970; New York: Touchstone, 1998).

Schenkel, Elmar, "Die verkehrte Insel: *The Tempest* und H.G. Wells's *The Island of Dr. Moreau*," *Anglia* 111 (1993), 39–58.

Schivelbusch, Wolfgang, *The Railway Journey: The Industrialization of Time and Space in the Nineteenth Century* (Berkeley: University of California Press, 1977).

Schore, Allan N., *Affect Regulation and the Repair of the Self* (New York: Norton, 2003).

Scott, Derek B., "The Sexual Politics of Victorian Musical Aesthetics," *The Journal of the Royal Musical Association* 119.1 (1994), 91–114.

Segal, Elizabeth, "The Gypsy Breynton Series: Setting the Pattern for American Tomboy Heroines," *Children's Literature Association Quarterly* 14 (Summer 1989), 61–71.

Showalter, Elaine, *The Female Malady: Women, Madness, and English Culture, 1830–1980* (New York: Pantheon Books, 1985).

——, *Hystories: Hysterical Epidemics and Modern Culture* (New York: Columbia University Press, 1997).

Shuttleworth, Sally, *Charlotte Brontë and Victorian Psychology* (Cambridge: Cambridge University Press, 1996).

Siegel, Bernie, *Love, Medicine and Miracles: Lessons Learned about Self-Healing from a Surgeon's Experience with Exceptional Patients* (1986; New York: HarperCollins, 2002).

Silverman, Debora L., *Art Nouveau in* Fin-de-Siècle *France: Politics, Psychology, and Style* (Berkeley: University of California Press, 1989).

Skerritt, Paul W., "Anxiety and the Heart: A Historical Review," *Psychological Medicine* 13 (1983), 17–25.

Slonimsky, Nicolas, *Lexicon of Musical Invective* (Seattle: University of Washington Press, 1965).

Smith, Andrew, *Victorian Demons: Medicine, Masculinity and the Gothic at the* Fin-de-Siècle (Manchester: Manchester University Press, 2004).

Smith, Roger, *Inhibition: History and Meaning in the Sciences of Mind and Brain* (Berkeley: University of California Press, 1992).

Snow, C.P., *The Two Cultures and A Second Look* (1963; New York: Cambridge University Press, 1979).

Stephan, Philip, *Paul Verlaine and the Decadence, 1882–90* (Manchester: Manchester University Press, 1974).

Stiles, Anne, "Cerebral Automatism, the Brain, and the Soul in Bram Stoker's *Dracula*," *Journal of the History of the Neurosciences* 15.2 (June 2006), 131–52.

——, "Robert Louis Stevenson's *Jekyll and Hyde* and the Double Brain," *Studies in English Literature, 1500–1900* 46.4 (Autumn 2006), 879–900.

Sulloway, Frank J., *Freud, Biologist of the Mind: Beyond the Psychoanalytic Legend* (New York: Basic Books, 1979).

Sutton, Emma, *Aubrey Beardsley and British Wagnerism in the 1890s* (Oxford: Oxford University Press, 2002).

Swenson, Kristine, *Medical Women and Victorian Fiction* (Columbia: University of Missouri Press, 2005).

Talbott, John, *Mind Wounds: War and Psychic Injury since 1860* (work in progress).

——, "Soldiers, Psychiatrists, and Combat Trauma," *Journal of Interdisciplinary History* 27 (Winter 1997), 437–54.

Thagard, Paul, *Hot Thought: Mechanisms and Applications of Emotional Cognition* (Cambridge: MIT Press, 2006).

Thrailkill, Jane F., *Affecting Fictions: Mind, Body, and Emotion in American Literary Realism* (Cambridge: Harvard University Press, 2007).

Tilton, Eleanor M., *Amiable Autocrat: A Biography of Dr. Oliver Wendell Holmes* (New York: Henry Schuman, 1947).

Trede, Fiona, "Traumhaft Schön? Madeleine G. und der Tanz im Unbewussten" *Tanzdrama* 58.3 (2001), 5–7.

Treitel, Corinna, *A Science for the Soul: Occultism and the Genesis of the German Modern* (Baltimore: Johns Hopkins University Press, 2004).

Vrettos, Athena, "Defining Habits: Dickens and the Psychology of Repetition," *Victorian Studies* 42 (Spring 1999/2000), 399–426.

Walker, Alan, *Franz Liszt*, 2 vols (Ithaca: Cornell University Press, 1987).

Weeks, Jeffrey, *Coming Out: Homosexual Politics in Britain, from the Nineteenth Century to the Present* (London: Quartet Books, 1977).
Wegener, Frederick, "'Few Things More Womanly or Noble': Elizabeth Stuart Phelps and the Advent of the Woman Doctor in America," *Legacy* 22.1 (2005), 1–17.
Weiner, Marc A., *Richard Wagner and the Anti-Semitic Imagination* (Lincoln: University of Nebraska Press, 1995).
Weliver, Phyllis, *Women Musicians in Victorian Fiction 1860–1900: Representations of Music, Science and Gender in the Leisured Home* (Aldershot: Ashgate, 2000).
West, Martin, *Ancient Greek Music* (Oxford: Clarendon Press, 1992).
White, Kevin, *The First Sexual Revolution: The Emergence of Male Heterosexuality in Modern America* (New York: New York University Press, 1993).
Will, Barbara, "The Nervous Origins of the American Western," *American Literature* 70 (June 1998), 293–316.
Wilson, Edmund, *Patriotic Gore: Studies in the Literature of the American Civil War* (New York: Oxford University Press, 1969).
Wolffram, Heather, "On the Borders of Science: Psychical Research and Parapsychology in Germany, c. 1870–1939" (Ph.D. diss., University of Queensland, 2005).
Wood, Jane, *Passion and Pathology in Victorian Fiction* (Oxford: Oxford University Press, 2001).
Young, Allan, *The Harmony of Illusions: Inventing Post-Traumatic Stress Disorder* (Princeton: Princeton University Press, 1995).
Young, Robert, *Mind, Brain, and Adaptation in the Nineteenth Century: Cerebral Localization and its Biological Context from Gall to Ferrier* (1970; New York: Oxford University Press, 1990).

Online sources

Carroll, Roz, "'Thinking through the Body': An Interview with Allan Schore," http://www.thinkbody.co.uk/papers/interview-with-allan-s.htm (accessed 21 January 2007).
Harrington, Ralph, "The Railway Accident: Trains, Trauma, and Technological Crisis in Nineteenth-Century Britain" (1999), http://www.york.ac.uk/inst/irs/irshome/papers/rlyacc.htm (accessed 24 January 2007).
Mitchell, Silas Weir, *Doctor and Patient* (1888), http://www.truthbeknown.com/doctor_and_patient.html (accessed 21 January 2007).
——, *Wear and Tear or, Hints for the Overworked* (1871), http://www.gutenberg.org/files/13197/13197-h/13197-h.htm (accessed 21 January 2007).

Index